Low-Level Hexavalent Chromium Treatment Options: Bench-Scale Evaluation

AF794565

Awwa Research Foundation

About the Awwa Research Foundation

The Awwa Research Foundation (AwwaRF) is a member-supported, international, nonprofit organization that sponsors research to enable water utilities, public health agencies, and other professionals to provide safe and affordable drinking water to consumers.

The Foundation's mission is to advance the science of water to improve the quality of life. To achieve this mission, the Foundation sponsors studies on all aspects of drinking water, including supply and resources, treatment, monitoring and analysis, distribution, management, and health effects. Funding for research is provided primarily by subscription payments from approximately 1,000 utilities, consulting firms, and manufacturers in North America and abroad. Additional funding comes from collaborative partnerships with other national and international organizations, allowing for resources to be leveraged, expertise to be shared, and broad-based knowledge to be developed and disseminated. Government funding serves as a third source of research dollars.

From its headquarters in Denver, Colorado, the Foundation's staff directs and supports the efforts of more than 800 volunteers who serve on the board of trustees and various committees. These volunteers represent many facets of the water industry, and contribute their expertise to select and monitor research studies that benefit the entire drinking water community.

The results of research are disseminated through a number of channels, including reports, the Web site, conferences, and periodicals.

For subscribers, the Foundation serves as a cooperative program in which water suppliers unite to pool their resources. By applying Foundation research findings, these water suppliers can save substantial costs and stay on the leading edge of drinking water science and technology. Since its inception, AwwaRF has supplied the water community with more than $300 million in applied research.

More information about the Foundation and how to become a subscriber is available on the Web at **www.awwarf.org**.

Low-Level Hexavalent Chromium Treatment Options: Bench-Scale Evaluation

Prepared by:
Philip Brandhuber, **Michelle Frey**, **Michael J. McGuire**, **PengFei Chao**, and **Chad Seidel**
McGuire Environmental Consultants, Inc., Denver, Colorado

Gary Amy and **Jaekyung Yoon**
University of Colorado, Boulder, Colorado

Laurie McNeill
Utah State University, Logan, Utah

and

Kashi Banerjee
North American Technology Center, US Filter Inc., Pittsburgh, Pennsylvania

Jointly sponsored by:
Awwa Research Foundation
6666 West Quincy Avenue, Denver, CO 80235-3098
The City of Los Angeles Department of Water and Power
The City of Glendale, California
The City of Burbank, California
The City of San Fernando, California
and
National Water Research Institute (NWRI)

Published by:

Awwa Research Foundation
Advancing the Science of Water®

IWA Publishing

DISCLAIMER

This study was funded by the Awwa Research Foundation (AwwaRF). AwwaRF assumes no responsibility for the content of the research study reported in this publication or for the opinions or statements of fact expressed in the report. The mention of trade names for commercial products does not represent or imply the approval or endorsement of AwwaRF. This report is presented solely for informational purposes.

Published by IWA Publishing, Alliance House, 12 Caxton Street, London SW1H 0QS, UK

Telephone: +44 (0) 20 7654 5500; Fax: +44 (0) 20 7654 5555; Email: publications@iwap.co.uk;
Web: **www.iwapublishing.com**

AwwaRF report number 91042F. Originally published by AwwaRF for its subscribers in 2004.
IWA Publishing version published 2005.

ISBN 1 84339 924 5

CONTENTS

LIST OF TABLES ix

LIST OF FIGURES xi

FOREWORD xvii

ACKNOWLEDGMENTS xix

EXECUTIVE SUMMARY xxi

CHAPTER 1: INTRODUCTION 1
- Research Needs 2
- Research Objectives 2
- Research Approach 3
- Research Team 3

CHAPTER 2: LITERATURE REVIEW 5
- Chromium Occurrence 5
- Health Effects and Regulatory Status 6
- Overview of Chromium Chemistry 8
- Potential Treatment Technologies 10
  - Conventional Coagulation–Flocculation 10
  - Lime Softening 11
  - Activated Carbon 11
  - Ion Exchange 11
  - Reverse Osmosis, Nanofiltration, and Electrodialysis 12
  - Influence of Chlorine 12
- Summary 13

CHAPTER 3: TREATMENT PROCESSES, METHODS, AND MATERIALS 15
- Chromium Oxidation–Reduction Experiments (Chapter 5) 15
  - Treatment Chemistry 15
  - Process Description 16
  - Experimental Protocol 16
  - Conditions Tested 18
  - Chromium Analysis 18
- Adsorption Experiments (Chapter 6) 20
  - Process Chemistry 20
  - Process Description 21
  - Batch Experiments 22
  - Column Experiments 25
  - Exhausted Media Leaching And Digestion Experiment 27
- Anion Exchange Experiments (Chapter 6) 28
  - Treatment Chemistry 28
  - Process Description 29

Column Experiments 29
Reactor Experiments (MIEX® resin) 30
Membrane Technology Experiments (Chapter 7) 32
Treatment Chemistry 32
Process Description 32
Membranes 32
Water Supplies 34
Membrane Testing Units 36
Membrane Oxidative Damage 38
Reduction and Precipitation Experiments (Chapter 8) 40
Phase E—Cr(VI) Reduction Experiments In Glendale Water 40
Cr(VI) Reduction Experiments In Synthetic Water Supplies 41
Cr(VI) Reduction And Total Chromium Removal Experiments In Synthetic And Natural Waters 41

CHAPTER 4: CHROMIUM OCCURRENCE AND CO-OCCURRENCE ANALYSIS 43
Objectives of the Co-Occurrence Analysis 43
Source Data 43
Creating the Chromium Groundwater Co-Occurrence Database 44
Characteristics of Chromium Occurrence 45
Characteristics of Chromium Co-Occurrence 49
Comparison With Project Water Supplies 52
Summary 53

CHAPTER 5: CHROMIUM REDUCTION AND OXIDATION 55
Control Experiments 55
Trivalent Chromium Oxidation Results 59
Dissolved Oxygen 59
Free Chlorine 59
Chloramine 63
Potassium Permanganate 63
Hydrogen Peroxide 64
Hexavalent Chromium Reduction Results 70
Stannous Chloride 70
Reduction by Sulfide 73
Reduction by Sulfite 75
Reoxidation Experiment Results 77
Chlorine Reoxidation Experiment 77
Permanganate Reoxidation Experiment 77
Additional Hexavalent Chromium Reduction and Chloramine Reoxidation Results 79
Summary of Findings 80
Oxidation by Dissolved Oxygen 80
Oxidation by Free Chlorine 80
Oxidation by Chloramines 82
Oxidation by Permanganate 82
Oxidation by Hydrogen Peroxide 82
Reduction by Stannous Chloride 82
Reduction by Sulfide 83

Reduction by Sulfite ... 83

CHAPTER 6: SORPTION AND ANION EXCHANGE TECHNOLOGY ASSESSMENT ... 85
Batch Experiments ... 85
Initial Capacity Experiment ... 85
Refined Carbonate-Buffered Capacity Experiment ... 86
pH Sensitivity Experiment ... 89
Dual Solute Experiments ... 90
Continuous Flow Experiments ... 93
Ion Exchange ... 93
Sulfur-Modified Iron ... 97
Granular Ferric Hydroxide Column Experiment ... 103

CHAPTER 7: MEMBRANE TECHNOLOGY: BENCH-SCALE RESULTS OF CHROMATE, ARSENATE, AND PERCHLORATE REJECTION BY NEGATIVELY CHARGED MEMBRANES ... 105
Membrane Properties ... 105
Effect of pH and Conductivity on Zeta Potential ... 105
Rejection of Chromate, Arsenate, and Perchlorate Under Conditions of Varying pH, Conductivity, and Membrane Zeta Potential ... 108
Effect of Co- and Counter-Ion Rejection on Anion Transport in Synthetic Water ... 115
Effect of pH and Conductivity on AS(III) Rejection and Membrane Zeta Potential ... 115
Hexavalent Chromium Rejection and Flux Decline in Various Natural Water Supplies ... 125
Membrane Oxidative Damage by CR(VI) Species ... 126
Results of Long-Term Monitoring ... 126
Summary ... 137

CHAPTER 8: ASSESSMENT OF HEXAVALENT CHROMIUM TREATMENT BY IRON-INDUCED REDUCTION AND PRECIPITATION–COPRECIPITATION AND COAGULATION ... 139
Background ... 139
Phase A—Evaluation of Iron Reduction Kinetics ... 140
Phase B—Impact of Co-Occurring Parameters on Reduction Kinetics ... 142
Phase C—Total Chromium Removal in Synthetic Water ... 144
Phase D—Reduction Kinetics and Total Chromium Removal in Experiments With a Natural Water Source Supplied by LADWP ... 145
Phase E—Simulation of Total Chromium Removal in Experiments With a Natural Water Source Supplied by Glendale ... 150
Summary ... 154

CHAPTER 9: ASSESSMENT OF CHROMIUM TREATMENT TECHNOLOGIES ... 155
Summary of the Performance of Chromium Treatment Technologies ... 155
Recommendations for Pilot-Testing ... 157

APPENDIX A: CO-OCCURRING CONTAMINANTS IN WATER SUPPLIES WITH HIGH AND LOW CHROMIUM CONCENTRATIONS ... 159

APPENDIX B: CHROMIUM REDUCTION AND OXIDATION CHEMISTRY—SUPPLEMENTAL RESULTS ... 161

REFERENCES ... 171

ABBREVIATIONS ... 177

LIST OF TABLES

Table ES.1 Testing program for the partnership study on treatment options for low-level hexavalent chromium xxv

Table ES.2 Comparative performance of investigated technologies xxvi

Table 2.1 Detections of hexavalent chromium in California drinking water sources*† 7

Table 2.2 Various international standards for total chromium 7

Table 2.3 Effects of resin matrix and porosity on run length 12

Table 3.1 Quality of the synthetic water 19

Table 3.2 Oxidants and reductants tested 19

Table 3.3 Treatment media evaluated in batch adsorption experiments 22

Table 3.4 Dual solute concentrations evaluated in dual solute experiments 26

Table 3.5 Column experiments on sulfur-modified iron treatment media 26

Table 3.6 Sulfur-modified iron media sample sizes for the TCLP and WET 28

Table 3.7 Properties of ion exchange resins used in column experiments on conventional anion exchange 30

Table 3.8 Characteristics of the membranes tested 34

Table 3.9 Quality parameters of the synthetic and natural water supplies tested 35

Table 3.10 Amounts of salts used to adjust conductivity 35

Table 3.11 Ion concentrations in feed waters 36

Table 3.12 Summary of the experimental plan for reduction and precipitation studies 40

Table 4.1 Database parameters and data acceptance criteria 46

Table 4.2 Characteristics of groundwater supplies containing total chromium (in mg/L, alkalinity in mg/L as $CaCO_3$, Si as mg/L SiO_2) 50

Table 4.3 Characteristics of groundwater supplies containing hexavalent chromium (in mg/L, alkalinity in mg/L as $CaCO_3$, Si as mg/L SiO_2) 50

Table 4.4 Degree of correlation between chromium and co-occurring parameters, as measured by Pearson Correlation Coefficients at a 95% confidence level 52

Table 4.5 Comparison of project water supplies and nationwide U.S. water supplies with total chromium concentrations of ≥10 μg/L 53

Table 5.1 Conditions tested in redox experiments 55

Table 5.2 Effect of potassium permanganate on measured Cr(VI) 56

Table 6.1 Summary of results from refined carbonate-buffered capacity experiment 88

Table 6.2 Results of pH sensitivity experiment 89

Table 6.3 Characterization of dual solute effects* on hexavalent chromium removal 92

Table 6.4 Average percent removal of dual solutes by treatment media 93

Table 6.5 Summary of column experiments on sulfur-modified iron treatment media 101

Table 6.6 Results of leachate testing in sulfur-modified iron columns 103

Table 7.1 Cr(VI) rejection with each combination of membrane and water source 125

Table 7.2 Effect of CT on normalized flux (J/J_o) as a function of pH (%) 130

Table 7.3 Effect of CT on normalized NaCl rejection as a function of pH (%) 133

Table 8.1 Water quality characteristics of LADWP well AT002 147

Table 8.2 Chromium removal from LADWP water by reduction and precipitation with an Fe:Cr weight ratio of 10:1 150

Table 8.3 Glendale treatment plant water quality (fall 2003) 152

Table 9.1 Performance of Cr(VI) treatment technologies, based on bench-scale testing 157

LIST OF FIGURES

Figure ES.1 Potential pathways for chromium control by drinking water utilities xxiii

Figure 2.1 Most commonly present metals in all matrices at Superfund sites 6

Figure 2.2 (A) pE–pH diagram of aqueous chromium; (B) Cr(III) solubility at 25°C 9

Figure 2.3 Relative distribution of Cr(VI) species in water as a function of pH and Cr(VI) concentration 9

Figure 2.4 Chemistry of chromium in coagulation–flocculation processes 10

Figure 3.1 Treatment options involving the reduction of Cr(VI) to Cr(III)............................... 17

Figure 3.2 Hexavalent chromium treatment by physical adsorption 20

Figure 3.3 Hexavalent chromium treatment by chelation... 20

Figure 3.4 Hypothesized mechanism for hexavalent chromium treatment by surface reduction and precipitation... 21

Figure 3.5 Treatment by adsorption or chelation.. 21

Figure 3.6 Protocol for batch adsorption experiments... 24

Figure 3.7 Conventional ion exchange process 29

Figure 3.8 Continuously mixed ion exchange process (MIEX®) 31

Figure 3.9 Small-scale MIEX® pilot unit.. 31

Figure 3.10 Reverse osmosis–nanofiltration process... 33

Figure 3.11 Membrane array.. 33

Figure 3.12 Schematic of stirred-cell filtration unit... 37

Figure 3.13 Schematic of experimental flat-sheet test cell 37

Figure 3.14 Distribution of Cr(VI) species and experimental conditions as functions of pH and concentrations 39

Figure 4.1 Database development flow chart 45

Figure 4.2 Cumulative probability distribution for total and hexavalent chromium in U.S. groundwater supplies 47

Figure 4.3 Total chromium occurrence in the contiguous United States 48

Figure 4.4 Hexavalent chromium occurrence in the contiguous United States 48

Figure 4.5 Cumulative probability distributions of pH in water supplies containing low and elevated concentrations of total chromium 51

Figure 5.1 Control experiment with 100 μg/L Cr(VI) in synthetic water 57

Figure 5.2 Control experiment with 100 μg/L Cr(III) in synthetic water (all measurements were for total chromium) 58

Figure 5.3 Control experiment with 100 μg/L Cr(III) in modified synthetic water (all measurements were for total chromium) 58

Figure 5.4 Control experiment with 10, 20, and 40 μg/L Cr(III) (all measurements were for total chromium) 59

Figure 5.5 Oxidation of 100 μg/L Cr(III) by 1 mg/L Cl_2 in deionized water 60

Figure 5.6 Oxidation of 100 μg/L Cr(III) by 1 mg/L Cl_2 in synthetic water 61

Figure 5.7 Oxidation of 100 μg/L Cr(III) by 1 mg/L Cl_2 in synthetic water at pH 9 61

Figure 5.8 Oxidation of 90 μg/L Cr(III) by 1 mg/L Cl_2 in Glendale water 62

Figure 5.9 Oxidation of 90 μg/L Cr(III) by 1 mg/L Cl_2 in Glendale water at pH 9 62

Figure 5.10 Oxidation of 100 μg/L Cr(III) by 1 mg/L Cl_2 in the reducing water at pH 5 65

Figure 5.11 Oxidation of 100 μg/L Cr(III) by 1 mg/L Cl_2 in the reducing water at pH 7 65

Figure 5.12 Oxidation of 100 μg/L Cr(III) by 0.5 mg/L chloramine in synthetic water 66

Figure 5.13 Oxidation of 100 μg/L Cr(III) by 0.5 mg/L chloramine in Glendale water at pH 7. 66

Figure 5.14 Oxidation of 100 μg/L Cr(III) by 1 mg/L $KMnO_4$ in deionized water 67

Figure 5.15 Oxidation of 100 μg/L Cr(III) by 1 mg/L $KMnO_4$ in synthetic water 67

Figure 5.16 Oxidation of 100 μg/L Cr(III) by $KMnO_4$ in synthetic water at pH 9 (samples shown are unfiltered) 68

Figure 5.17 Oxidation of 100 μg/L Cr(III) by 1 mg/L $KMnO_4$ in synthetic water at pH 9 68

Figure 5.18 Oxidation of 100 μg/L Cr(III) by 3 mg/L $KMnO_4$ in Glendale water 69

Figure 5.19 Oxidation of 100 μg/L Cr(III) by 3 mg/L $KMnO_4$ in Glendale water at pH 9 69

Figure 5.20 Reduction of 100 μg/L Cr(VI) by 1.3 mg/L $SnCl_2$ in deionized water (samples shown are unfiltered) 70

Figure 5.21 Reduction of 100 μg/L Cr(VI) by 1.3 mg/L $SnCl_2$ in synthetic water at pH 5........ 71

Figure 5.22 Reduction of 100 μg/L Cr(VI) by 1.3 mg/L $SnCl_2$ in synthetic water at pH 7........ 72

Figure 5.23 Reduction of 100 μg/L Cr(VI) by 1.3 mg/L $SnCl_2$ in synthetic water at pH 9........ 72

Figure 5.24 Reduction of 100 μg/L Cr(VI) by 1.3 mg/L $SnCl_2$ in modified synthetic water...... 73

Figure 5.25 Reduction of 100 μg/L Cr(VI) by 1 mg/L sulfide in deionized water........ 74

Figure 5.26 Reduction of 100 μg/L Cr(VI) by 1 mg/L sulfide in synthetic water........ 74

Figure 5.27 Reduction of 100 μg/L Cr(VI) by 1 mg/L sulfite in synthetic water........ 75

Figure 5.28 Reduction of 100 μg/L Cr(VI) by 10 mg/L sulfite in synthetic water........ 76

Figure 5.29 Reduction of 100 μg/L Cr(VI) by 10 mg/L sulfite in Glendale water........ 76

Figure 5.30 Control test for reoxidation experiment—reduction of 100 μg/L Cr(VI) by ferrous iron in Glendale water at pH 7........ 78

Figure 5.31 Reduction of 100 μg/L Cr(VI), followed by reoxidation by 1 mg/L chlorine 78

Figure 5.32 Reoxidation of 100 μg/L Cr(VI) by 3 mg/L permanganate 79

Figure 5.33 Combined chlorine (chloramine) concentrations over the experimental period 81

Figure 5.34 Cr(III) reoxidation to Cr(VI) with different chloramine doses 81

Figure 6.1 Results of initial capacity experiment—24-hr samples........ 85

Figure 6.2 Results of refined carbonate-buffered capacity experiment using a 1-g/L media dose—2-hr samples........ 86

Figure 6.3 Increased Cr(VI) removal with increased media doses after 2 hours........ 87

Figure 6.4 Effect of time on Cr(VI) removal in experiments using a 1-g/L media dose........ 87

Figure 6.5 Results of pH sensitivity experiment using a 1-g/L media dose—2-hr samples........ 89

Figure 6.6 Effect of sulfate on hexavalent chromium removal 91

Figure 6.7 Effect of bicarbonate on hexavalent chromium removal 91

Figure 6.8 Breakthrough curves for strong base anion exchange resins........ 94

Figure 6.9 Comparison of resin performance 96

Figure 6.10 MIEX® Experimental results from small-scale pilot unit 97

Figure 6.11 Results from initial capacity column experiment—sulfur modified iron media 98

Figure 6.12 Results of column experiment on pH effect—sulfur-modified iron media 99

Figure 6.13 Results of confirmatory pH effect column experiment—sulfur-modified iron media 100

Figure 6.14 Results of batch isotherm tests with sulfur-modified iron treatment media 102

Figure 6.15 Results of total digestion of sulfur-modified iron treatment media 102

Figure 6.16 Results of column experiment with granular ferric hydroxide treatment media for removal of co-occurning solutes—Cr(VI) and arsenic 104

Figure 7.1 Effect of pH on the zeta potential of RO, NF, and UF membranes 106

Figure 7.2 Effect of conductivity on the zeta potential of RO, NF, and UF membranes 107

Figure 7.3 Effect of pH on membrane rejection of chromate, arsenate, and perchlorate at a solution conductivity of 30 mS/m (conductivity adjusted with KCl) 109

Figure 7.4 Effect of conductivity on membrane rejection of chromate, arsenate, and perchlorate at pH 8 (conductivity adjusted with KCl) 111

Figure 7.5 Effect of electrolytes on membrane rejection of chromate, arsenate, and perchlorate at pH 8 and a solution conductivity of 30 mS/m 113

Figure 7.6 Effect of pH on membrane rejection of co- and counter-ions at a solution conductivity of 30 mS/m 116

Figure 7.7 Effect of conductivity on membrane rejection of co- and counter-ions at pH 8 120

Figure 7.8 Effect of pH and conductivity on membrane rejection of As(III) 124

Figure 7.9 Rejection of Cr (VI) and other parameters in natural water sources 127

Figure 7.10 Flux decline in RO, NF, and tight UF membranes with natural water sources 128

Figure 7.11 FTIR spectra of clean and damaged GM membranes (pH 4, CT = 8.84×10^6 mg-hr/L 136

Figure 8.1 The chromium reduction and precipitation/coagulation process 140

Figure 8.2 Comparison of Cr(VI) reduction by Fe(II) and Fe(0) in synthetic water 141

Figure 8.3 Impact of co-occurring contaminants sulfate, phosphate, arsenate, and silica on Cr(VI) reduction by Fe(II) and Fe(0) ... 143

Figure 8.4 Comparison of total chromium removal in synthetic water by 0.45-µm and 0.2-µm filters at an Fe(II):Cr(VI) dosage of 10:1 ... 145

Figure 8.5 Comparison of total chromium removal in synthetic water with various filter sizes and Fe(II):Cr(VI) weight ratios of 10:1, 50:1, and 100:1 ... 146

Figure 8.6 Removal of Cr(VI) and total chromium using the reduction–precipitation process in LADWP water ... 148

Figure 8.7 Comparison of chromium speciation in LADWP water after reduction and 0.2-µm filtration ... 149

Figure 8.8 MINEQL simulation of Fe(II)-complexed species as a function of pH in water from LADWP well AT002 ... 151

Figure 8.9 Possible inhibition pathway for Cr(VI) reduction by Fe(II) and coprecipitation or coagulation of Cr(II) in the presence of silica ... 151

Figure 8.10 Comparison of filtered total chromium concentrations after reduction with Fe(II) in Glendale water spiked to 100 µg/L Cr(VI) ... 153

Figure 8.11 Influence of polymer addition on filtered total chromium concentration (pH = 6.5) ... 153

Figure 9.1 Alternative methods for controlling chromium ... 155

Figure A.1 Comparison of co-occurring contaminants in high- and low-chromium water supplies ... 159

Figure B.1 Control experiment for deionized water spiked with 100 µg/L Cr(VI) (total chromium was not measured) ... 161

Figure B.2 Control experiment for Glendale water spiked with 100 µg/L Cr(VI) (concentrations of filtered and unfiltered samples were the same) ... 162

Figure B.3 Control experiment for modified synthetic water spiked with 100 µg/L Cr(VI)—total chromium and Cr(VI) concentrations were the same (data not shown) ... 162

Figure B.4 Control experiment for deionized water spiked with 100 µg/L Cr(III) ... 163

Figure B.5 Control experiment for reducing water spiked with 100 µg/L Cr(III) (no Cr(VI) was detected in any sample) ... 163

Figure B.6 Control experiment for Glendale water spiked with 100 µg/L Cr(III) ... 164

Figure B.7 Oxidation of 100 μg/L Cr(III) by 1 mg/L Cl_2 in synthetic water at pH 5................ 164

Figure B.8 Oxidation of 100 μg/L Cr(III) by 1 mg/L Cl_2 in synthetic water at pH 7................ 165

Figure B.9 Oxidation of 100 μg/L Cr(III) by 1 mg/L chloramine in deionized water.............. 165

Figure B.10 Oxidation of 100 μg/L Cr(III) by potassium permanganate in synthetic water at pH 5 (all samples are unfiltered)... 166

Figure B.11 Oxidation of 100 μg/L Cr(III) by potassium permanganate in synthetic water at pH 7 (all samples are unfiltered)... 166

Figure B.12 Oxidation of 100 μg/L Cr(III) by 3 mg/L potassium permanganate in reducing water at pH 5.. 167

Figure B.13 Oxidation of 100 μg/L Cr(III) by 3 mg/L potassium permanganate in reducing water at pH 9.. 167

Figure B.14 Reduction of 100 μg/L Cr(VI) by 1.3 mg/L stannous chloride in Glendale water at pH 5.. 168

Figure B.15 Reduction of 100 μg/L Cr(VI) by 1.3 mg/L stannous chloride in Glendale water at pH 7.. 168

Figure B.16 Reduction of 100 μg/L Cr(VI) by 1.3 mg/L stannous chloride in Glendale water at pH 9.. 169

FOREWORD

The Awwa Research Foundation (AwwaRF) is a nonprofit corporation that is dedicated to the implementation of a research effort to help utilities respond to regulatory requirements and traditional high-priority concerns of the industry. The research agenda is developed through a process of consultation with subscribers and drinking water professionals. Under the umbrella of a Strategic Research Plan, the Research Advisory Council prioritizes the suggested projects based upon current and future needs, applicability, and past work; the recommendations are forwarded to the Board of Trustees for final selection. The foundation also sponsors research projects through the unsolicited proposal process; the Collaborative Research, Research Applications, and Tailored Collaboration programs; and various joint research efforts with organizations such as the U.S. Environmental Protection Agency, the U.S. Bureau of Reclamation, and the Association of California Water Agencies.

This publication is a result of one of these sponsored studies, and it is hoped that its findings will be applied in communities throughout the world. The following report serves not only as a means of communication of the results of the water industry's centralized research program but also as a tool to enlist the further support of nonmember utilities and individuals.

Projects are managed closely from their inception to the final report by the foundation's staff and large cadre of volunteers who willingly contribute their time and expertise. The foundation serves a planning and management function and awards contracts to other institutions such as water utilities, universities, and engineering firms. The funding for this research effort comes primarily from the Subscription Program, through which water utilities subscribe to the research program and make an annual payment proportionate to the volume of water they deliver and consultants and manufacturers subscribe based on their annual billings. The program offers a cost-effective and fair method for funding research in the public interest.

A broad spectrum of water supply issues is addressed by the foundation's research agenda: resources, treatment and operations, distribution and storage, water quality and analysis, toxicology, economics, and management. The ultimate purpose of the coordinated effort is to assist water suppliers in providing the highest possible quality of water economically and reliably. The true benefits are realized when the results are implemented at the utility level. The foundation's trustees are pleased to offer this publication as a contribution toward that end.

Hexavalent chromium may enter water supplies through naturally occurring sources as well as anthropogenic sources such as electroplating, tanning and textile manufacturing. Recent health assessments conducted in the state of California and reaction to the popular movie *Erin Brockovich* have sensitized the public to potential, but as yet unsubstantiated, hazards of chromium in drinking water. Though the current maximum contaminant level (MCL) for total chromium established by the U.S. Environmental Protection Agency is 100 μg/L, the state of California is in the process of developing an MCL specifically for hexavalent chromium. At present, the treatment of chromium in drinking water is rare. The objectives of this work were to better understand the occurrence of chromium, investigate chromium oxidation and reduction properties under drinking water conditions, and determine the feasibility of treating hexavalent chromium to very low concentrations. Results of this research will assist drinking water utilities in meeting more stringent requirements for the treatment of chromium.

Walter J. Bishop
Chair, Board of Trustees
Awwa Research Foundation

James F. Manwaring, P.E.
Executive Director
Awwa Research Foundation

ACKNOWLEDGMENTS

This report was made possible by the efforts of many people. At the inevitable risk of omitting someone, particular thanks go to the following people for their contributions to the project:

AwwaRF Project Advisory Committee
Hakan Gurleyuk, Frontier Geosciences
David Kimbrough, Castaic Lake Water Agency
Bruce Macler, U.S. Environmental Protection Agency
Richard Nagel, West Basin Metropolitan Water District

McGuire Environmental Consultants
Michael MacPhee
Nicole Blute
Shereen Lanzarotta

Los Angeles Department of Water and Power
Pankaj Parekh
Lucik Melikian

City of Glendale Water and Power
Donald Froelich
Miriam Sykes

University of Colorado
R. Scott Summers
Scott Davis

Utah State University
Han Lai
Julia Rahman
Brian Thomas
Crystal Viator

EXECUTIVE SUMMARY

Many water utilities throughout the United States have found contamination of their water supplies by hexavalent chromium [Cr(VI)]. Chromium occurs naturally in many water sources, depending on the hydrogeology of the region. Water sources can also be affected by Cr(VI) contamination plumes from industrial centers, landfills, and improper discharge of industrial processing streams. Though the presence of total chromium in drinking water supplies is limited under current standards (the federal limit is 0.10 mg/L and the state of California's limit is 0.050 mg/L), concerns have been raised regarding exposure to low concentrations of Cr(VI) through drinking water.

Scientists have long recognized the disparity of concerns about chromium toxicity based on its chemical speciation. Trivalent chromium [Cr(III)] is considered a micronutrient and does not present any potential health threat at the concentrations found in ambient water environments. However, Cr(VI) has been demonstrated to be a human carcinogen when inhaled. The health effects of Cr(VI) through ingestion—the dominant exposure route for drinking water—have had limited study, and these studies have yielded uncertain conclusions. Because of the uncertainties about Cr(VI) health effects, utilities and public health officials have begun to investigate the feasibility of limiting Cr(VI) concentrations in drinking water to very low concentrations (<2 μg/L).

The water utilities of the California cities of Glendale, Burbank, Los Angeles, and San Fernando recognized that controlling Cr(VI) in drinking water supplies may be required. An initial study commissioned by the city of Glendale found little or no available information on the performance of various techniques for reducing Cr(VI) to very low concentrations. Industrial or hazardous waste treatment applications for Cr(VI) have long been established, but the performance of these technologies with chromium concentrations typically found in drinking water had never been determined. The city of Glendale, at the request of Congressman Adam Schiff, commissioned a three-phase study to develop technology for removing trace levels of Cr(VI) from drinking water supplies. This $3 million effort consisted of Phase 1 bench-scale studies, Phase 2 pilot-scale studies, and Phase 3 demonstration-scale studies. Congressman Schiff and U.S. Senator Barbara Boxer initiated efforts at the federal level to fund these studies. The Los Angeles Department of Water and Power in conjunction with Glendale Water and Power, Burbank Water and Power, and the city of San Fernando formed a partnership with the American Water Works Association Research Foundation (AwwaRF) and the National Water Research Institute (NWRI) to conduct the Phase 1 bench-scale tests to investigate the feasibility of low-level Cr(VI) control in drinking water.

The purpose of this report is to present the results of this partnership study, which included an analysis of chromium occurrence and co-occurrence, an evaluation of Cr(VI) removal technologies, and an examination of chromium oxidation and reduction chemistry. The removal technologies evaluated included sorption processes using various granular or resin media; anion exchange using a range of commercially available resins; membrane treatment using reverse osmosis, nanofiltration, and ultrafiltration membranes; and conventional treatment by Cr(VI) reduction and precipitation with a coagulant. Each of these technologies was investigated at bench scale to assess which ones showed the most promise for controlling low-level Cr(VI) in drinking water. Subsequent research on larger scale application of the more promising technologies could be based on the findings of this study.

CHROMIUM REDUCTION AND OXIDATION CHEMISTRY

The stability of chromium species in drinking water is an important concern for utilities trying to achieve very low Cr(VI) concentrations in the water delivered to consumers' taps. The difference in the toxicity of Cr(VI) (possible human carcinogen) and Cr(III) (micronutrient) could result in Cr(VI) reduction as a viable control strategy for utilities. The kinetics and effectiveness of Cr(VI) reduction to Cr(III) must be known in order to design appropriate systems to effect this shift in speciation. Further, the stability of Cr(III) in the presence of oxidants potentially present in or applied to drinking water supplies must be understood in order to assess the likely success of such control strategies.

Reduction of Cr(VI) to Cr(III) by stannous chloride and ferrous iron compounds was highly effective. Stannous chloride achieved complete reduction of Cr(VI) to Cr(III) within minutes, as determined by the study's first samples, collected at 30 min. However, previous researchers found that stannous chloride could reduce Cr(VI) to Cr(III) in a matter of seconds and that the completeness of the reduction depended on the applied weight ratio of stannous chloride to Cr(VI) (McGuire Environmental Consultants 2002). Reduction of Cr(VI) by sulfide was slow and incomplete, whereas reduction by sulfite was rapid but also incomplete (40% reduction at a 1,000:1 weight ratio of sulfite to Cr(VI)). Reduction of Cr(VI) by ferrous iron was fairly rapid; complete reduction was achieved within 1 to 2 hr. Therefore, in terms of applied dosage requirements and rate of reduction kinetics, stannous chloride was the most efficient and effective reductant evaluated in this study.

Oxidation of soluble Cr(III) to Cr(VI) is likely to take place as a result of the addition of treatment chemicals such as disinfectants (free chlorine or chloramines), potassium permanganate, or hydrogen peroxide. However, particulate Cr(III) species, such as those formed from reduction of Cr(VI) by ferrous iron, could not be oxidized by any of the oxidants tested. Naturally occurring oxidants are not expected to contribute to the oxidation of Cr(III) to Cr(VI). Dissolved oxygen was ineffective at oxidizing Cr(III), and other naturally occurring compounds that could serve as electron donors would be expected to be less effective than dissolved oxygen. Free chlorine, however, oxidized 50–65% of the Cr(III) to Cr(VI) at neutral or low pH conditions. Measurable and significant Cr(III) oxidation was observed when monochloramine was applied (0.5 to 10 mg/L total chlorine dose). Potassium permanganate was found to be a very effective oxidant of Cr(III); nearly complete oxidation occurred within 60 min at neutral or low pH conditions. Thus, utilities considering a Cr(VI) reduction strategy should keep in mind that the presence of free chlorine, monochloramine or potassium permanganate could adversely affect the success of this approach.

The effect of pH was important to the evaluation of chromium oxidation and reduction. Lower pH conditions were preferred in completing and increasing the reaction rates for both the reduction of Cr(VI) (except in the case of sulfite where pH effects were not observed) and the oxidation of Cr(III). In the presence of elevated concentrations of Cr(III) (100 µg/L), high pH conditions (pH 9) resulted in substantial precipitation of Cr(III) species. The particulate Cr(III) was not oxidized by any chlorine-based oxidant but did appear to be oxidized by potassium permanganate.

COMPARATIVE ASSESSMENT OF LOW-LEVEL CR(VI) REMOVAL TECHNOLOGIES

The ability to control chromium in drinking water has several potential pathways (Figure ES.1). This study investigated nearly all of the potential methods of controlling Cr(VI) either through the use of technologies that remove Cr(VI) directly (adsorption, anion exchange, membrane filtration, electrocoagulation) or those that remove the reduced form of chromium, Cr(III) (precipitation with membranes or coagulation and precipitation with conventional or membrane filters).

These technologies were investigated using laboratory-scale testing methods, including batch isotherm tests, bench membrane systems, flow-through mini-columns, and jar testing techniques. The testing procedures and general conditions used for each technology in the study are described in Table ES.1. The tests were conducted at the participating laboratories of the University of Colorado (Boulder) and Utah State University (Logan). For most technologies, controlled water matrices were used to screen performance. Selected technologies were further assessed using natural groundwater matrices from the Los Angeles Department of Water and Power and the Glendale Water and Power groundwater facilities.

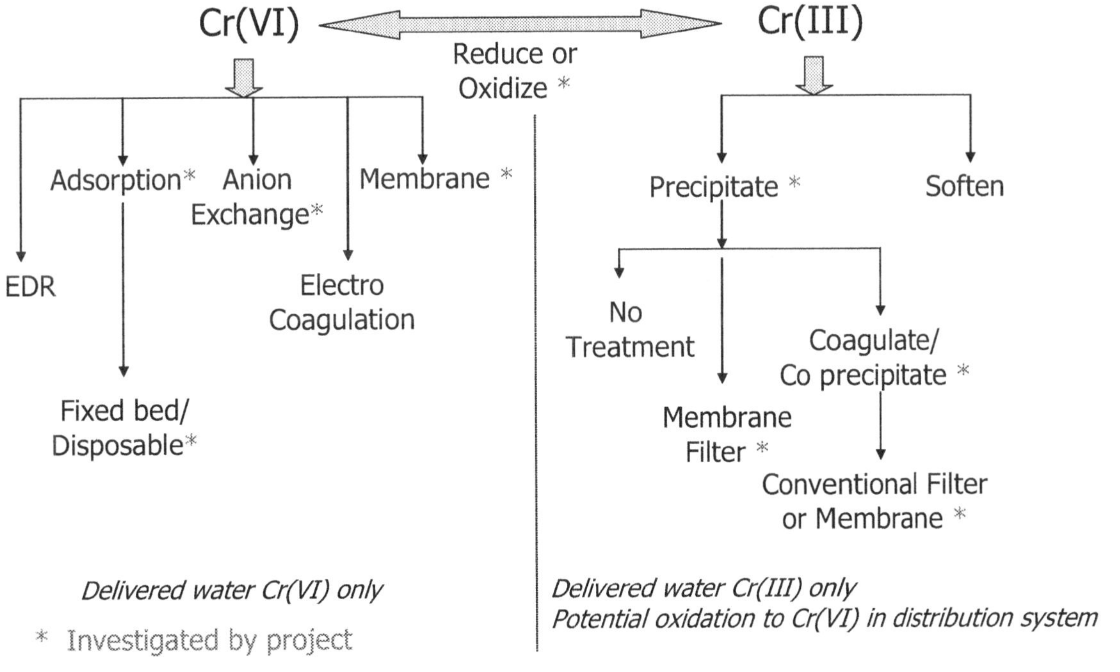

Figure ES.1 Potential pathways for chromium control by drinking water utilities

At the outset of the study, the research team expected sorptive media that have been effective at arsenate removal to also be effective at removing Cr(VI), but this was not the case in the bench-scale investigation (Table ES.2). Indeed, the most effective sorptive media for arsenate (granular ferric hydroxide and granular ferric oxide) both performed poorly at removing Cr(VI) from water. Sulfur-modified iron, however, was extremely effective at Cr(VI) removal and is a disposable media that can effectively control arsenate as well. Among the other media evaluated, only the iron-coated activated alumina, iron-impregnated zeolite, and chelating media (metal-binding ligand media) showed good Cr(VI) removal potential. Upon testing with at least one other solute present in deionized (DI) water, though, only the sulfur-modified iron media appeared to be robust through a range of pH conditions and competitive or interference effects from co-occurring constituents.

The anion exchange resins tested all provided excellent removal of Cr(VI) in modified DI water matrices. Among the anion exchange resins, the magnetic resin was tested in a similar fashion to the sorptive media because it is applied as a dispersed particulate in a mixing chamber and then removed through sedimentation. In this case, contact time becomes important to the effectiveness of the resin. Application of fixed-bed anion exchange resins does not involve the same kind of kinetic effect because all of the water comes into direct contact with the resin immediately. Therefore, for fixed-bed resins, the empty bed contact time (EBCT) will determine the capacity of the resin (run length) rather than reflect the kinetics of Cr(VI) exchange.

On a mass basis, all of the anion exchange resins demonstrated significantly higher Cr(VI) removal capacities than any of the sorptive media. The run length exhibited by the resins (10% breakthrough of a 1-mg/L Cr(VI) influent stream at 1,000 to 2,000 bed volumes and 2.5 min EBCT) implies substantially longer productivity potential than observed for arsenate. Indeed, depending on the source water Cr(VI) concentrations and the targeted effluent concentrations, anion exchange resins can achieve many more bed volumes of productivity for Cr(VI) removal than for arsenate removal.

Membrane removal of Cr(VI) was effective at bench scale under a wide range of pH and conductivity conditions. The research demonstrated that membrane charge can strongly affect Cr(VI) rejection performance, and a more negatively charged membrane is desirable. When uncharged membranes are used, more alkaline pH conditions can improve Cr(VI) rejection. However, increasing conductivity has a strong impact on the performance of ultrafiltration and notable effects on nanofiltration performance. Reverse osmosis membranes were only negligibly affected by increasing conductivity.

Coagulation techniques for controlling chromium can be effective when coupled with reduction of Cr(VI) to Cr(III), sufficient mass of coagulant or coagulant aid polymer, and control of coagulation pH. Conditions tested in this study for reducing chromium in treated water (highest ratio of ferrous iron to Cr(VI), lowest pH, and use of a coagulant aid polymer) did not reduce the treated water chromium concentration below 2 µg/L when initial Cr(VI) concentrations were either 100 µg/L or 50 µg/L. This includes samples filtered through a 1,000-dalton filter. These results suggest that Cr(III) solubility may control the effectiveness of chromium removal by this technology. However, the potential for more complete removal of Cr(III) using higher ferrous iron doses needs to be tested at pilot scale. The fate of the reduced Cr(III) in the presence of residual chlorine or chloramines in utility distribution systems would be the critical factor in assessing whether unacceptable levels of Cr(VI) would be delivered to consumers' taps.

Table ES.1
Testing program for the partnership study on treatment options for low-level hexavalent chromium

Technology class	Testing program component		
	Media	Water matrices	Type of bench-scale test
Sorption/chelation	Eleven granular or resin-based media	DI water with single or dual co-occurring solutes and Cr(VI) spike; natural water matrices from LADWP and Glendale groundwater supplies with Cr(VI) spike; flow-through mini-column tests with Cr(VI)-spiked potable water for a limited number of media	All media tested by batch isotherm procedures (total Cr(VI) removal capacity); three media evaluated using flow-through mini-column tests
Ion exchange	One magnetized strong base anion exchange resin; four strong base anion exchange resins—two type I resins versus two type II resins	DI water with single or dual co-occurring solutes and Cr(VI) spike, and local potable water spiked with Cr(VI) (magnetized strong base anion exchange resin only); DI water with background levels of bicarbonate and sulfate with Cr(VI) spike (4 remaining resins)	Magnetized strong base anion exchange resin evaluated by batch isotherm procedures and flow-through bench-scale unit; all remaining conventional resins evaluated using flow-through mini-column tests
Membranes	Five membranes (one ultrafiltration membrane, two nanofiltration membranes, and two reverse osmosis membranes	Modified DI water using potassium chloride salt spiked with Cr(VI), arsenate, and perchlorate; natural water matrices from LADWP and Glendale groundwater supplies with Cr(VI) spike	Bench-top membrane (flat sheet or fiber) units in a batch operational mode
Coagulation with precipitation	Variety of stoichiometric ratios of ferrous iron to Cr(VI) applied with and without coagulation aids	DI water modified with potassium chloride and bicarbonate with Cr(VI) spike; dual co-occurring constituent tests on modified DI water including silicate, sulfate, phosphate, arsenate and pH; natural water matrices from LADWP and Glendale groundwater supplies with Cr(VI) spike	Jar test experiments with five filter sizes applied to the settled water: 0.45 µm, 0.2 µm, 100,000 daltons, 10,000 daltons, and 1,000 daltons

Table ES.2
Comparative performance of investigated technologies

Media/Membrane type	Technology class	Cr(VI) treatment performance		Observations on technology performance	Selected for additional tests
		Removal*	Kinetics†		
		Sorption and chelation media			
GFH	Granular ferric hydroxide	Poor to good	Some effect	Low pH strongly preferred; interference from SO_4, HCO_3, and TOC observed; co-removal of arsenate feasible	Dual solute: YES Mini-column: YES
Bayoxide E33	Granular ferric oxide	Poor	Some effect	Low pH strongly preferred; interference from SO_4, HCO_3, and TOC observed; co-removal of arsenate feasible	Dual solute: YES Mini-column: NO
Sulfur-modified iron	Sulfur-modified iron	Excellent	No effect	No pH efforts or interferences observed	Dual solute: YES Mini-column: YES
Alcan AA-400	Activated alumina	Poor	No effect	Not available	Dual solute: NO Mini-column: NO
Alcan AA-F550	Iron-coated activated alumina	Good	Some effect	No pH effect	Dual solute: YES Mini-column: NO
WTR Z24	Iron-impregnated zeolite	Good	Strong effect	Excellent removal at 24 hr; low pH preferred; interference with Cl, SO_4, and TOC observed	Dual solute: YES Mini-column: NO
ADI-G2	Iron oxide–coated diatomite	Poor	No effect	Neutral pH preferred	Dual solute: YES (pH only) Mini-column: NO
Vintec Bauxsol	Bauxite clay	Poor	No effect	Not available	Dual solute: NO Mini-column: NO
Metre General Octolig	Metal-binding ligand	Good	No effect	Low pH strongly preferred; interference from SO_4, HCO_3, and TOC observed	Dual solute: YES Mini-column: NO

(continued)

Table ES.2
Comparative performance of investigated technologies

Media/Membrane type	Technology class	Cr(VI) treatment performance		Observations on technology performance	Selected for additional tests
		Removal*	Kinetics†		
Anion exchange resins					
Orica MIEX®	Magnetic anion exchange	Excellent	Some effect	No pH effect; interference from SO_4; co-removal of arsenate, nitrate, perchlorate, and TOC	Dual solute and bench-scale flow-through tests
Rohm & Haas Amberlite IRA410	Strong base type II	Excellent	No effect	Cr(VI) removal strongly preferred over sulfate removal	Mini-column test only
DOWEX Marathon MSA	Strong base type I	Excellent	No effect	Cr(VI) removal strongly preferred over sulfate removal	Mini-column test only
Purolite A-600	Strong base type I	Excellent	No effect	Cr(VI) removal strongly preferred over sulfate removal	Mini-column test only
Purolite A-300	Strong base type II	Excellent	No effect	Cr(VI) removal strongly preferred over sulfate removal	Mini-column test only
Membrane technologies					
GM	Ultrafilter	Poor	No effect	No pH effect; strong interference by conductivity	NA
MX07	Nanofilter	Excellent	No effect	Neutral to alkaline pH preferred; some interference by conductivity	NA
LFC-1	Reverse osmosis	Excellent	No effect	Neutral to alkaline pH preferred; negligible interference by conductivity	NA
TFC-ULP	Reverse osmosis	Good	No effect	Tested in natural water only	NA
TFC-S	Nanofilter	Good	No effect	Tested in natural water only	NA

(continued)

Table ES.2
Comparative performance of investigated technologies

Media/Membrane type	Technology class	Cr(VI) treatment performance		Observations on technology performance	Selected for additional tests
		Removal*	Kinetics†		
Coagulation and precipitation technologies					
Ferrous sulfate coagulant	Cr(VI) reduction to Cr(III) with precipitation and filtration with and without coagulant aid polymer	Good	Some effect	Low pH strongly preferred in natural water; 100:1 weight ratio of ferrous iron to Cr(III) could be necessary for effective chromium removal; limited efficiency by Cr(III) solubility	NA

* Cr(VI) Removal performance based on 24-hr equilibrium concentration with media dosage of 1 g/L

- Excellent = >90% removal
- Good = 50%–90% removal
- Poor = <50% removal

† Kinetic effect shows improved Cr(VI) removal from 30 min to 2 hr and 24 hr

- Strong = >50% removal or reduction improvement
- Weak = 10%–50% removal or reduction improvement
- None = <10% removal or reduction improvement

NA = not applicable

RECOMMENDED TECHNOLOGIES FOR LARGER SCALE EVALUATIONS

The screening of numerous technologies for low-level chromium control has identified the technologies that are most promising for application by water utilities. Subsequent study of the technologies listed below is necessary to characterize their performance under larger scale flow-through conditions:

- **Anion exchange** (both as fixed-bed and dispersed-contactor applications). The effect of regeneration on anion exchange performance and the feasibility of operating anion exchange on a "throw-away" basis are issues that can be addressed in the next phase of study.
- **Sulfur-modified iron sorption media**. The mechanism of Cr(VI) control by this technology needs to be better understood—Cr(VI) sorption or Cr(VI) reduction to Cr(III) with removal by the iron precipitates. Post-treatment requirements for iron control also need to be investigated.
- **Membrane treatment by nanofiltration and reverse osmosis**. Rejection of Cr(VI) by membrane technologies is excellent, but the operational conditions (productivity, fouling, membrane life) that affect the feasibility of this technology may require further evaluation. Membrane treatment will not be carried forward to pilot-scale studies because of the large loss of water associated with this technology.

- **Coagulation and precipitation of reduced Cr(III)**. Reduction and removal of chromium as Cr(III) can achieve reasonably low concentrations of chromium in treated water. In combination with a post-treatment and disinfectant regime, this strategy could prove extremely effective for controlling Cr(VI) at very low concentrations. Further study of mixing conditions and process kinetics is required to optimize this process for reducing Cr(VI) to very low concentrations.

Drinking water providers face new challenges in controlling Cr(VI) in potable water supplies. This study has identified several technologies that are likely to be highly effective at controlling chromium in drinking water supplies. Future research needs can be prioritized on the basis of the results of this study. In particular, further work is needed to determine which of the successful technologies is the most cost-effective. These studies will enable the drinking water community to understand Cr(VI) treatment and control.

CHAPTER 1
INTRODUCTION

In February 1999 the California Office of Environmental Health Hazard Assessment (OEHHA) issued a Public Health Goal (PHG) for total chromium of 2.5 µg/L (OEHHA 1999). The PHG, based on a 10^{-6} risk level for 0.2 µg/L hexavalent chromium [Cr(VI)], was 40 times less than the U.S. Environmental Protection Agency's (USEPA's) maximum contaminant level (MCL) of 100 µg/L for total chromium [Cr(III) + Cr(VI)]. According to OEHHA, "Chromium VI is the most toxic form of chromium and is the primary health concern when chromium is present in drinking water." The success of the movie *Erin Brockovich*, which popularized a groundwater chromium pollution lawsuit in Hinkley, Calif., has sensitized the public to the health hazards of chromium in drinking water. Public pressure has led to several governmental actions at both the federal and state levels.

- The National Toxicology Program announced in May 2001 that it would assess the carcinogenicity of Cr(VI).
- U.S. Senators Harry Reid (Nevada) and Barbara Boxer (California) sponsored a bill requiring USEPA to establish a federal standard for Cr(VI) in drinking water. The bill was never signed into law.
- The California state legislature in 2001 passed a bill (SB 351—Ortiz) requiring the DHS to adopt an MCL for Cr(VI).
- The California Environmental Protection Agency requested the University of California to establish a blue ribbon panel to review scientific questions concerning the potential of Cr(VI) to cause cancer when ingested.
- The California Department of Health and Human Services (DHS) requested OEHHA to establish a specific Public Health Goal (PHG) for Cr(VI). Establishing a PHG is the first step toward developing a state drinking water standard specifically for Cr(VI). (Although the PHG for total chromium was issued in 1999, it was withdrawn in November 2001.)

All of these actions must be viewed from the perspective that, at the time, no technology had been demonstrated to be effective at treating chromium to concentrations consistent with the total chromium PHG or the Cr(VI) 10^{-6} risk level of 0.2 µg/L.

Chromium VI is a contaminant that:

- May not currently be regulated sufficiently to protect human health,
- Could be regulated to a very low level in the future,
- Evokes consumer concern,
- May be present on a nationwide scale, and
- Has no proven treatment that can reduce it to very low concentrations.

The challenge to drinking water utilities is clear. Thus, the goal of this project was to investigate the feasibility of reducing Cr(VI) to very low concentrations in drinking water supplies, consistent with California's actions.

RESEARCH NEEDS

A 2001 study conducted on behalf of Glendale Water and Power and the cities of Los Angeles, San Fernando, and Burbank concluded that no treatment technology had been demonstrated to produce finished drinking water with Cr(VI) concentrations consistent with California's health risk level of 0.2 μg/L (McGuire Environmental Consultants 2001). In fact, the treatment of chromium in drinking water is rare. Chromium treatment has been limited primarily to applications such as industrial waste streams, groundwater remediation, and cooling tower water to which Cr(VI) was added as a corrosion inhibiter. Several technologies have the ability to treat high chromium concentrations to relatively high effluent concentrations (<50 μg/L) and have the potential to reduce Cr(VI) to very low concentrations. Possible technologies for removing low-level Cr(VI) include:

- Ion exchange media;
- Reduction of Cr(VI) to Cr(III) followed by precipitation, coagulation, and filtration;
- Membranes; and
- Sorption-based metal oxide media or chelation media.

Unfortunately, the performance of all of these treatment technologies may be inherently limited due to various factors. For example, Cr(III) solubility may be sufficiently high to render the reductive precipitation process ineffective at controlling very low concentrations of total chromium, or competing ions may limit the effectiveness of Cr(VI) removal by sorptive media. A fundamental evaluation of the performance of these and other removal technologies was required to determine the feasibility of treating low concentrations of chromium. In addition, because Cr(VI) is listed as a hazardous waste under the U.S. Resource Conservation and Recovery Act, each technology must be evaluated with respect to its residual stream. The quantity, character, and stability of the residual must be examined for each treatment process.

As indicated, chromium has a complex acid–base and oxidation–reduction (redox) chemistry. Both the effectiveness of chromium treatment technologies and chromium's potential health effects are related to speciation. Fundamental research into the kinetics of chromium oxidation and reduction reactions, the stability of speciation, and the influence of co-occurring contaminants on chromium redox and equilibrium chemistry is required. For example, the fate of Cr(III) entering a distribution system in the presence of chlorine needs to be examined because even small amounts of Cr(III) oxidizing to Cr(VI) could potentially raise Cr(VI) levels above concentrations of concern.

Given the drinking water industry's limited experience with chromium control, pilot-scale testing is required to develop baseline experience with treating low-level Cr(VI). This is particularly important because pilot testing can assist with characterizing residuals. Like arsenic treatment, the handling and disposal of potentially toxic chromium residuals may be more complex than the treatment process itself.

RESEARCH OBJECTIVES

The objectives of the research program were to:

- Gain a better understanding of chromium occurrence;
- Determine the feasibility of treating low-level Cr(VI), using bench-scale systems;
- Compare the performance of alternative Cr(VI) treatment technologies;

- Quantify the chemistry and kinetics of chromium oxidation and reduction processes under water quality conditions typically found in drinking water treatment; and
- Make specific recommendations regarding appropriate pilot-scale technologies for subsequent testing at Glendale, Calif.

RESEARCH APPROACH

In keeping with the research objectives, a multiple-phase research program was developed. The program was divided into several major tasks, and the results of each task are presented as separate chapters in this report. The major tasks were to:

- Review existing literature (Chapter 2);
- Estimate chromium occurrence and co-occurrence (Chapter 4);
- Investigate chromium oxidation and reduction chemistry (Chapter 5);
- Evaluate Cr(VI) treatment technologies
  - Adsorption-based—including ion exchange (Chapter 6),
  - Membrane-based (Chapter 7),
  - Reduction-based (Chapter 8); and
- Assess technologies for treating Cr(VI) (Chapter 9).

RESEARCH TEAM

McGuire Environmental Consultants, Inc. (MEC) assembled a team to execute the project. The project team consisted of MEC, the University of Colorado (CU), Utah State University (USU), and the North American Technology Center of US Filter (NATC). The Los Angeles Department of Water and Power (LADWP) and the city of Glendale, Calif., provided assistance to the project team. Phil Brandhuber at MEC was principal investigator for the overall project as well as lead investigator for the literature review, the co-occurrence analysis, and the assessment of sorption and ion exchange technologies. Gary Amy at CU was responsible for the membrane technology assessment. Laurie McNeill at USU was responsible for the chromium oxidation and reduction experiments. Kashi Banerjee of NATC, with input from CU and MEC, was responsible for assessing iron-based Cr(VI) reduction; these experiments were conducted at CU. Michael McGuire was responsible for the final rewrite and edit of the report and for initiating two follow-up studies.

CHAPTER 2
LITERATURE REVIEW

CHROMIUM OCCURRENCE

Chromium, the twenty-first most abundant element in the earth's crust, is an important industrial metal used in diverse products and processes (Nriagu 1988). The principal ore of chromium is chromite ($FeCr_2O_4$) (Weast, Astle, and Beyer 1988). In the United States, chromium ore has been imported since 1961 and is refined into chromium metal and alloys (ATSDR 1993). Chromium has been used in industry for various purposes such as electroplating, leather tanning, and textile manufacturing, as well as controlling biofilms in cooling towers. It has been released into the environment through inadequate storage during manufacturing and improper waste disposal practices (Palmer and Wittbrodt 1991, Calder 1988). The National Priority List of 1986 developed by USEPA showed that about 40% of the 1,000 sites listed had reported metals problems. According to USEPA data (USEPA 1996a) for locations where Records of Decision have been signed, chromium is the second most common metal found at Superfund sites (Figure 2.1).

In addition to entering the environment through anthropogenic sources, chromium is also a natural constituent of groundwater supplies (Robertson 1975), rivers (Cox and McLeod 1992), and freshwater lakes (Kaczynski and Kieber 1993). Naturally occurring chromium results primarily from weathering of chromite and other chromium-bearing minerals present in bedrock and soils (Nriagu and Nieboer 1988). Chromium exists in water in either the trivalent Cr(III) or hexavalent Cr(VI) form. River systems in the United States have been found to have chromium concentrations ranging from less than 1 µg/L to 30 µg/L. Lakes in the United States usually contain less than 5 µg/L of chromium (ATSDR 1993). An earlier survey of U.S. drinking water sources conducted in 1974–1975 found chromium concentrations ranging from 0.4 to 8 µg/L, with a mean of 1.8 µg/L (ATSDR 1993). Water monitoring data collected by the California Department of Health Services from 1984 to 1997 show that total chromium was detected in 822 of 9,604 drinking water sources, or approximately 9% of the sources surveyed. Based on the practical detection limit of 10 µg/L, total chromium concentrations in the samples in which chromium was detected ranged from 10 µg/L to a maximum of 1,100 µg/L, with a mean of 23 µg/L.

An ongoing Cr(VI) sampling program conducted by the California DHS indicates that as of Dec. 1, 2003, 33% of the 6,229 drinking water sources monitored contained detectable concentrations of Cr(VI). Table 2.1 summarizes the DHS findings.

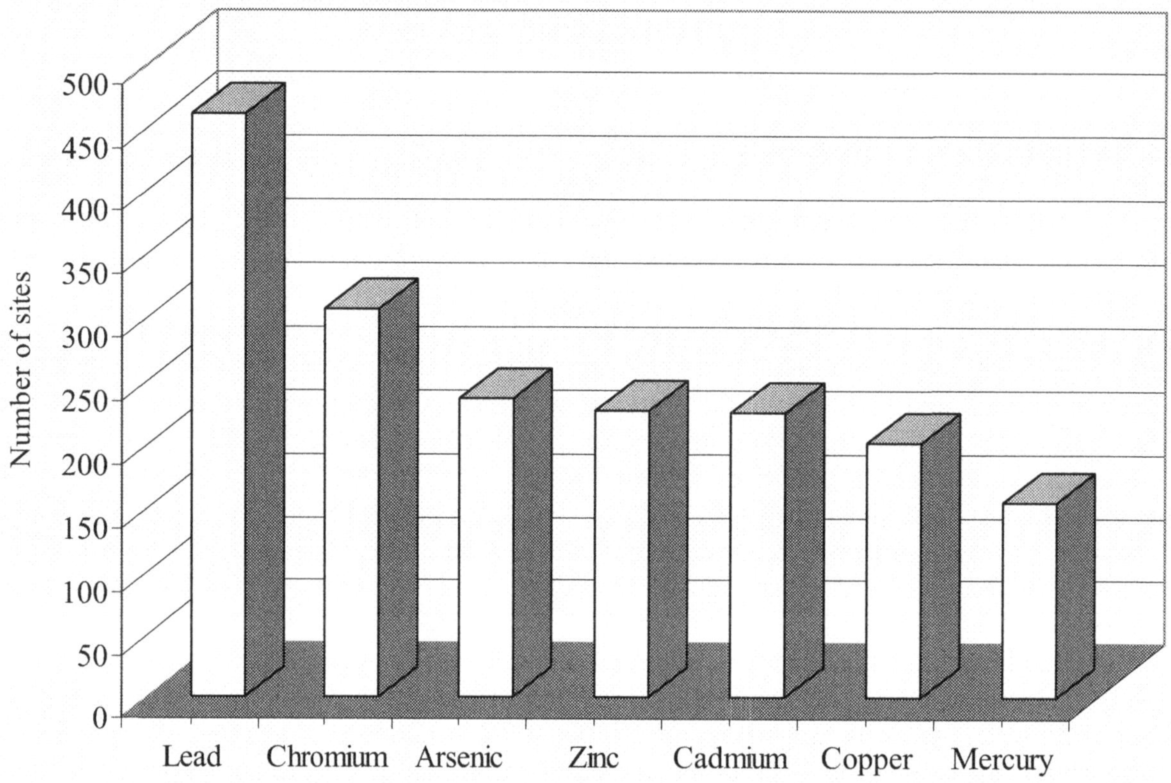

Figure 2.1 Most commonly present metals in all matrices at Superfund sites

HEALTH EFFECTS AND REGULATORY STATUS

Chromium first appeared in the U.S. Public Health Service's (USPHS's) Drinking Water Standards in 1946 because several studies showed that it readily induced skin sensitizations and that the hexavalent form produced lung tumors when inhaled (Machle and Gregorious 1942, USPHS 1953, Kimbrough 2002). Although adverse health effects were associated with inhalation of Cr(VI), the amount of chromium that can be ingested over a lifetime without posing health risks is still under evaluation (Flegal et al 2001). Nevertheless, chromium has beneficial as well as detrimental properties. Although Cr(III) is essential for human nutrition, especially in glucose metabolism (Vincent 2000), most hexavalent chromium compounds are toxic. In fact, several can cause lung cancer. USEPA set an MCL for total chromium in drinking water—Cr(III) plus Cr(VI)—because analytical methods at the time did not distinguish between the valence states (Sorg 1979). Despite the fact that Cr(III) is not considered toxic, the conservative assumption for protecting public health is that a substance that is carcinogenic by one route may be carcinogenic by other routes. This assumption has led the World Health Organization, USEPA, and other regulatory agencies to regulate total chromium (Table 2.2).

As noted in Chapter 1, the California OEHHA in February 1999 issued a Public Health Goal (PHG) for total chromium of 2.5 μg/L (OEHHA 1999). This PHG, based on a 10^{-6} risk level of 0.2 μg/L Cr(VI), was 40 times less than USEPA's MCL of 100 μg/L for total chromium.

OEHHA withdrew the PHG in November 2001 because it had been calculated with data from a study that OEHHA no longer plans to use in assessing health risks from chromium. Although a PHG for a chemical contaminant in drinking water does not necessarily indicate a significant risk to public health, California law requires that drinking water standards be set as close to the PHG as is economically and technically feasible. Therefore, OEHHA and DHS committed in March 2002 to develop the first PHG and drinking-water standard for Cr(VI) in the United States. Although California legislation specified that a Cr(VI) MCL was to be issued by Jan. 1, 2004, this deadline was not met, and a new date for issuing the MCL had not been set when this report went to press.

Table 2.1
Detections of hexavalent chromium in California drinking water sources*†

Peak concentration (μg/L)	Number of sources	Percentage of sources %
>50	5	-
46–50	2	-
41–45	3	-
36–40	5	-
31–35	7	-
26–30	20	-
21–25	22	-
16–20	59	1
11–15	134	2
6–10	406	7
1–5	1,346	22
Total Cr(VI) detections	2,068	33
Sources with no data or <1 μg/L	4,161	67
Total sources reporting	6,229	100

* As of Dec. 1, 2003

† Cr(VI) concentrations are from sources reporting more than a single detection. Sources may include both untreated and treated supplies, distribution systems, blending reservoirs, and other sampled entities. Table does not include agricultural wells, monitoring wells, or more than one representation of the same source (e.g., a source with data for both untreated and treated supplies is counted as a single source). Detections lower than the detection limit for reporting purposes are not included. Data should be considered draft.

Table 2.2
Various international standards for total chromium

Issuing organization	MCL (μg/L)	MCLG (μg/L)
California Code of Regulations	50	-
European Community	50	-
Health Canada	50	-
USEPA	100	100
World Health Organization	50	-

OVERVIEW OF CHROMIUM CHEMISTRY

Chromium, a metallic element with an atomic number of 24, is a member of periodic table group VIB along with molybdenum and tungsten. Chromium has four naturally occurring isotopes, but none of them is radioactive (Weast, Astle, and Beyer 1988). Chromium forms a number of salts, which are characterized by a variety of colors, solubilities, and other properties. The most important chromium salts are sodium and potassium chromates and dichromates, plus the potassium and ammonium chrome alums (Hodgman, Weast, and Selby 1961). Chromium has several oxidation states (Figure 2.2A), the most common and stable of which are Cr(II), Cr(III) and Cr(VI) (Baes and Mesmer 1976). Several Cr(IV) and Cr(V) species are known as intermediates in redox reactions and are unstable with respect to disproportionation to Cr (III) and Cr(VI).

In aquatic systems most chromium occurs in two oxidation states, Cr(III) and Cr(VI) (Figure 2.2B). Cr(III) occurs as a cation, and the hydroxide complex is insoluble. Cr(VI) occurs as an anion as either chromate ($HCrO_4^-/CrO_4^{2-}$) or dichromate ($Cr_2O_7^{2-}$). Both anionic forms are very soluble, and the formation of each is pH-dependent (Sengupta, Clifford, and Subramonian 1986). The simple ionic form of Cr(III) is $Cr(III)^+$, which predominates at pH <4. At pH >4, $Cr(III)^+$ forms hydroxide complexes in a stepwise fashion as pH increases ($Cr^{3+} \rightarrow Cr(OH)^{2+} \rightarrow Cr(OH)_2^+ \rightarrow Cr(OH)_3^0 \rightarrow Cr(OH)_4^-$). These complexation reactions control the ionic state of aqueous Cr(III), with the ionic charge changing from +3 to –1 between pH 4 and pH 10. At a pH range of 6–8, which is typical for natural water supplies, the predominant aqueous species is $Cr(OH)_3^0$. However, at this pH range Cr(III) exhibits low solubility, which is controlled by $Cr(OH)_3(s)$ (K_{SP} of $\approx 10^{-30}$). The minimum Cr(III) solubility is at pH ≈8. In contrast to Cr(III), Cr(VI) is highly soluble. At low concentrations, Cr(VI) is present in water as diprotic chromic acid (H_2CrO_4, $pK_{a1} = 0.81$, $pK_{a2} = 6.49$) (Butler 1967, Tong and King 1953). Thus, in natural water supplies two Cr(VI) oxyanion species predominate, monovalent $HCrO_4^-$ below pH 6.5 and divalent CrO_4^{2-} above pH 6.5 (Figure 2.3). An additional Cr(VI) species, dichromate ($Cr_2O_7^{2-}$), predominates at concentrations greater than 1,000 mg/L. It is unlikely that any drinking water source would contain such a high Cr(VI) concentration. However, it is possible that Cr(VI) concentrations at the surface of a treatment media may be high enough to favor the presence of dichromate.

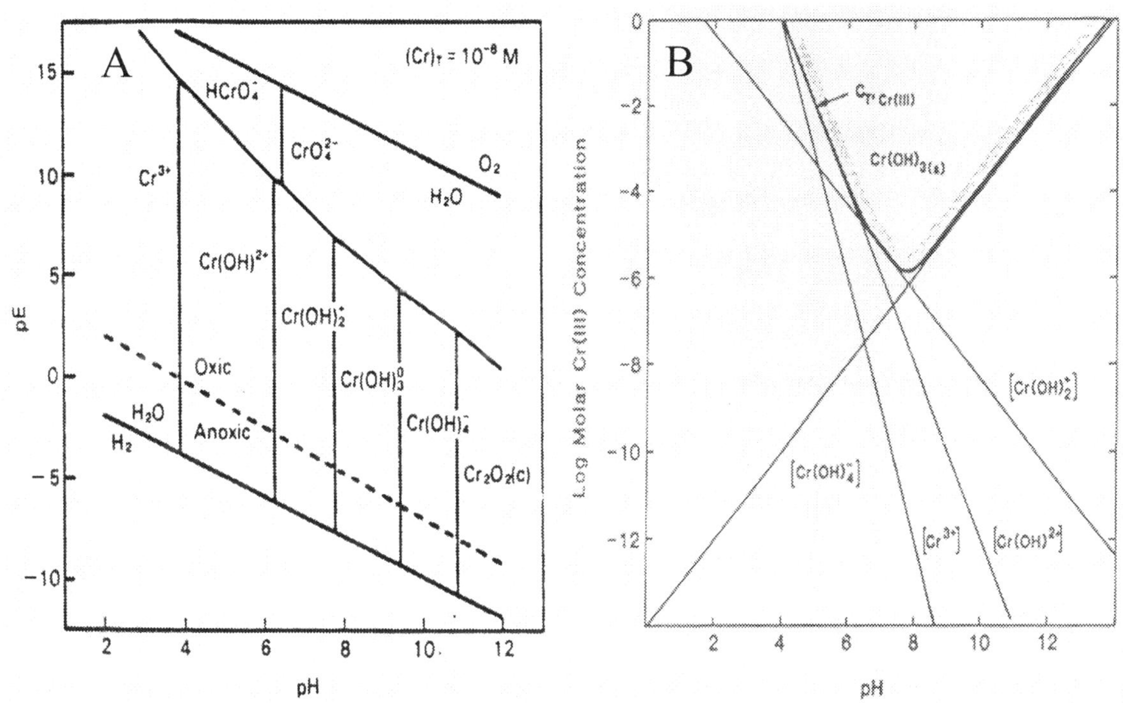

Figure 2.2 (A) pE–pH diagram of aqueous chromium; (B) Cr(III) solubility at 25°C

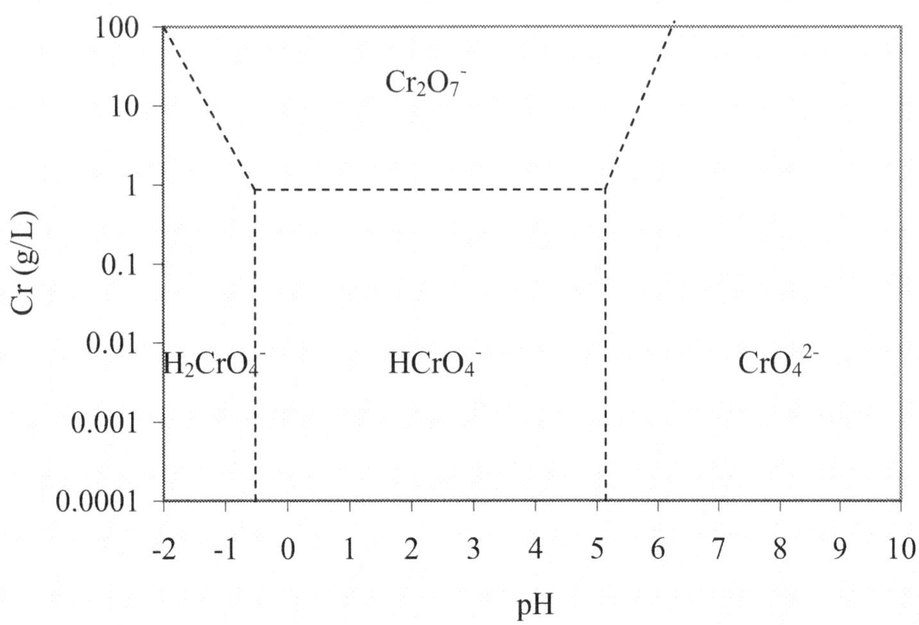

Figure 2.3 Relative distribution of Cr(VI) species in water as a function of pH and Cr(VI) concentration

POTENTIAL TREATMENT TECHNOLOGIES

Because chromium is used either directly or indirectly in many manufacturing processes, treatment techniques to remove or recover chromium from industrial wastewater have been well developed and documented (Sorg 1979, Patterson 1985). The mechanisms of chromium removal by means of conventional treatment technologies such as chemical reduction, coagulation, lime softening, filtration, and activated carbon adsorption include redox transformation, precipitation–dissolution, and adsorption–desorption reactions. These treatment technologies efficiently remove high influent concentrations (milligram-per-liter levels) of Cr(III), Cr(VI), or both in industrial chromium wastes (Patterson 1985). However, few studies have investigated treatment of the low chromium concentrations present in drinking water sources.

Viewed at a fundamental level, chromium's complex aqueous chemistry is advantageous because it exhibits several characteristics that can be exploited for treatment purposes. The ionic nature of Cr(VI) makes it suitable for treatment by processes that are driven by electrostatic attraction (sorption) or ionic replacement (ion exchange). Its metallic behavior makes it suitable for treatment by processes that form complexes that immobilize chromium by chelation. The low solubility of Cr(III) at pH ≈8 offers the possibility of treatment by precipitation. The relatively high standard electrode potential of chromic acid (E^0 = +1.195 V) implies that a number of reductants are likely to be effective at reducing Cr(VI) to Cr(III). And even though Cr(VI) is a relative small oxyanion (MW = 116 daltons), its divalent negative charge at pH >6.5 is highly suited to physiochemical separation by charged polymeric membranes. Possible treatment technologies for removing low chromium concentrations (microgram-per-liter levels) are summarized under the next five subheads.

Conventional Coagulation–Flocculation

The literature indicates that removal of Cr(III) by conventional coagulation and flocculation followed by filtration varies greatly according the nature of the source water and the coagulant used (Philipot, Chaffange, and Sibony 1984) (Figure 2.4). Cr(VI), however, is not removed by conventional treatment processes using alum or ferric sulfate as coagulants because the chromate and dichromate ions are extremely soluble (Sorg 1979). Effective removal by hydroxide precipitation requires that the Cr(VI) first be reduced to Cr(III) using a chemical reducing agent such as ferrous sulfate, sodium bisulfite, or stannous chloride (Besselievre 1969).

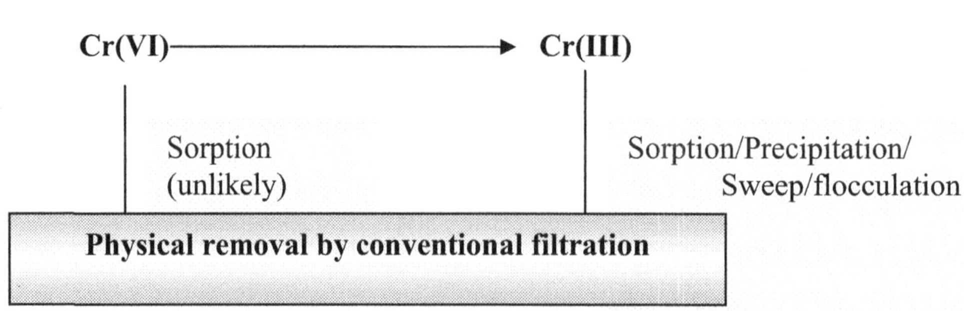

Figure 2.4 Chemistry of chromium in coagulation–flocculation processes

Lime Softening

In a similar manner to conventional coagulation, lime softening can effectively remove Cr(III), but removals for Cr(IV) are poor. The addition of reductants is needed to reduce Cr(VI) (Sorg 1979). However, the rate and efficiency of Cr(VI) reduction are favored at low or neutral pH conditions. High pH during softening processes hinders Cr(VI) reduction reactions. In addition, even though Cr(VI) can be reduced to Cr(III) and then removed by precipitation, residual solid $Cr(OH)_3$ might be re-dissolved to form aqueous Cr(III). Incomplete removal of Cr(III), therefore, can allow reoxidation to Cr(VI) in the presence of oxidants such as chlorine, chloramines, or ozone (Clifford and Chau 1987, Ulmer 1986). Thus, achieving chromium concentrations of less than 10 μg/L by using precipitative lime softening would be difficult.

Activated Carbon

It has been reported that activated carbon can remove high concentrations (milligram-per-liter levels) of chromium (Sorg 1979; Philipot et al. 1984; Han, Schlautman, and Batchelor 2000). So far, however, no literature has reported on the efficiency of removing low chromium concentrations from drinking water. Laboratory and pilot-plant studies conducted by Hung and Wu (1977) using powdered activated carbon (PAC) and granular activated carbon (GAC) showed that some Cr(VI) removal occurred at pH 2, and the efficiency increased with increasing pH (from pH 2 to pH 5–6). Yoshida and colleagues (1977) studied the adsorption of 10 mg/L of Cr(III) and Cr(VI) onto PAC and GAC as a function of pH. Both types of activated carbon readily adsorbed Cr(VI) in the pH range of 4–6.5 as the $HCrO_4^-$ and CrO_4^{2-} ions. Cr(III), however, was not adsorbed by GAC at any pH, and in acid solution Cr(VI) was easily reduced to Cr(III) in the presence of activated carbon. In addition, Linstedt and co-workers (1971) speculated that the higher Cr(VI) removals achieved by GAC might result from direct chemical interaction between the Cr(VI) and organic substances in the wastewater and subsequent removal of the organics onto the carbon.

Ion Exchange

Experience with treating industrial waste indicates that ion exchange can be economical for recovering chromium and eliminating waste discharge (Jakobsen and Laska 1977). Because ion exchange effectively removes high concentrations of chromium in wastewater, this process may be applicable for removing chromium from drinking water. Because of the positive charge of Cr(III) and the negative charge of Cr(VI), a two-step ion exchange process in which use of a cation resin for Cr(III) removal is followed by use of an anion resin for Cr(VI) removal could be effective if both species were present. However, because high removal efficiencies occur using extreme acid cation resins and strong base anion resins, this approach might not be practical for removing chromium from drinking water supplies.

Although all anion exchange resins strongly prefer Cr(VI), the resin matrix itself significantly influences run length to Cr(VI) breakthrough. Table 2.3 shows the effects of resin matrix and porosity on run length to Cr(VI) breakthrough (Clifford 1990). The longest run length (32,000 bed volumes and 98 days) was achieved with macroporous resins. A study conducted by Bahowick, Dobie, and Kumamoto (1996) demonstrated that ion exchange could reduce groundwater Cr(VI) concentrations averaging 30 μg/L to below detection limits (2 μg/L). The authors concluded that competing anions, especially sulfate and carbonate, had a significant impact on ion exchange performance in Cr(VI) removal. Clifford (1990) found the range of chromate recovery for 1N sodium chloride (NaCl) regenerations was 75–90%. The recovery, which was always less than 100%, was attributed to Cr(VI) reduction to Cr(III) with subsequent precipitation of greenish solid $Cr(OH)_{3(s)}$ precipitant. These low recoveries may limit the economic effectiveness of ion exchange as a treatment technique for chromate.

Reverse Osmosis, Nanofiltration, and Electrodialysis

A number of studies have shown that reverse osmosis (RO) achieves excellent removal of Cr(III) and Cr(VI). Electrodialysis is comparable to RO and therefore should also be an appropriate technique for reducing both forms of chromium in drinking water. Mixon (1973) conducted bench-scale RO studies on the removal of Cr(VI) from a variety of potable water, wastewater, and brackish water supplies with varying chromium concentrations of 0.9–10 mg/L. Results showed that when chromium concentrations were less than 5 mg/L, chromium rejections were at least 90 percent, regardless of water or membrane types. Other studies conducted by Hindin and colleagues (1968) on untreated water supplies with different Cr(VI) concentrations (0.47, 5, and 50 mg/L) showed that chromium removal was typically greater than 94%. High concentrations of chromium decreased removal efficiencies only by 1-2%. Though it is anticipated that nanofiltration will effectively remove Cr(VI), little data have been published on this subject.

Influence of Chlorine

Cr(III) can be oxidized to Cr(VI) by free chlorine. The rate of oxidation is highly dependent on pH and chlorine dosage (Toyama, Osuga, and Maruyama 1978; Ulmer 1986; Clifford and Chau 1987). Toyama and colleagues (1978) studied the oxidation of an initial Cr(III) dose of 10 mg/L at various pH levels (from less than pH 1 to pH 13) and observed that Cr(III) oxidized to Cr(VI) most rapidly and completely at pH 7. Ulmer (1986) observed a similar trend with a lower initial Cr(III) concentration (0.45 mg/L); maximum Cr(VI) conversion occurred between pH 6 and 7. Clifford and Chau (1987) conducted chlorination experiments using Cr(III) spiking and observed that up to 90% of the Cr(III) had been converted to Cr(VI) after 100 hours. In water supplies containing total organic carbon (TOC), conversion of Cr(III) to Cr(VI) was slower (Clifford and Chau 1987). Because the pH of most treated water ranges from pH 6 to pH 9, conditions under which reaction rates are higher, oxidation of Cr(III) by chlorine in the distribution system is expected.

Table 2.3
Effects of resin matrix and porosity on run length

Resin	Matrix/porosity	Capacity (meq/mL)	Run length to 10 mg/L Cr(VI)	
			Bed volume	Days
IRA 900	Polystyrene DVB/MR	1.1	32,000	98
Dowex11	Polystyrene DVB/iso	1.2	20,700	68
IRA958	Polyacrylic DVB/MR	0.8	14,600	44

Source: Clifford, D. 1990. Ion Exchange and Inorganic Adsorption. In *Water Quality and Treatment: A Handbook of Community Water Supplies*. Edited by F. Pontius. New York: American Water Works Association, McGraw-Hill, Inc. Reproduced with permission of The McGraw-Hill Companies.

SUMMARY

Chromium is an inorganic drinking water contaminant that has recently received considerable public attention. In aqueous systems, chromium exists in two oxidation states: the nontoxic trivalent form, Cr(III), and the potentially carcinogenic hexavalent form, Cr(VI). Although total chromium—Cr(III) plus Cr(VI)—has been monitored in drinking water systems for decades, few data exist concerning its speciation and its occurrence at concentrations below the current MCL. It is widely accepted that Cr(VI) is a human carcinogen via inhalation. However, at this time, there is no compelling evidence that ingestion of low levels of Cr(VI) is harmful to human health. The general principles of aqueous chromium chemistry are well understood, but the redox behavior of chromium species in the presence of chemicals typically used in drinking water treatment is not well documented. Although chromium treatment in industrial waste streams has been widely investigated, few treatment studies have investigated the treatment of chromium to very low concentrations in drinking water.

CHAPTER 3
TREATMENT PROCESSES, METHODS, AND MATERIALS

This chapter summarizes the treatment processes, experimental methods, and procedures used for the various areas of focus in this study: chromium oxidation–reduction experiments, Cr(VI) treatment by sorption and ion exchange, Cr(VI) treatment by membrane technology, and Cr(VI) treatment by iron reduction and precipitation–coprecipitation or coagulation. Chromium oxidation–reduction experiments were conducted at Utah State University (USU) under the direction of USU. The sorption and ion exchange experiments were conducted at the University of Colorado (CU) under the direction of McGuire Environmental Consultants, Inc. (MEC). The membrane technology experiments were conducted at CU under the direction of CU. The iron reduction experiments were conducted at CU under the direction of the North American Technology Center of US Filter with input from MEC. Experiments conducted at USU used USU facilities and equipment. Experiments conducted at CU used CU facilities and equipment.

CHROMIUM OXIDATION–REDUCTION EXPERIMENTS (CHAPTER 5)

Treatment Chemistry

Reducing Cr(VI) to Cr(III) is relatively easy to accomplish. It involves providing a source of electrons (reductant) so that Cr(VI) can be reduced to Cr(III). Potential reductants include stannous chloride, sulfide, sulfite, and ferrous iron compounds. Evaluation of these reductants indicated that stannous chloride and ferrous iron are the most effective reductants for treating Cr(VI) in drinking water. The Cr(VI) reduction reactions involving ferrous iron and stannous chloride are shown in Equations 3.1 and 3.2.

Ferrous iron

$$3\ Fe(OH)_2 + CrO_4^{2-} + 4\ H_2O \Rightarrow 3\ Fe(OH)_3 + Cr(OH)_3 + 2\ OH^- \qquad (3.1)$$

Stannous chloride

$$2H^+ + 3/2\ Sn(OH)_2 + CrO_4^{2-} + 4\ H_2O \Rightarrow 3/2\ SnO_2 + Cr(OH)_3 + H_2O \qquad (3.2)$$

The kinetics of the stannous chloride reaction appears to be faster than that of the ferrous iron reaction, and smaller doses of stannous chloride, on a stoichiometric basis, are required to complete the reduction reaction compared with ferrous iron.

Following reduction, at moderate pHs, Cr(III) precipitates as insoluble $Cr(OH)_3$. However, the solubility of $Cr(OH)_3$ ($Ksp = 6.3 \times 10^{-31}$) is in the microgram-per-liter range. Insoluble $Cr(OH)_3$ may be removed from treated water by filtration. Coprecipitation or coagulation with ferric hydroxides will increase the apparent size of $Cr(OH)_3$ precipitates, making them more amenable to removal by filtration.

Process Description

Three different treatment processes involving Cr(VI) reduction are presented in Figure 3.1. The first process, Option A, consists of adding a reductant directly to the process stream. A sufficient amount of reductant is added to complete the reduction process but no more. Assuming that complete reduction occurs, all the chromium is speciated in the nonhazardous trivalent form, and the treated water is supplied to the distribution system without any attempt to remove the Cr(III). Option B is similar to Option A, except that the process is performed at near pH 8—the minimum pH for $Cr(OH)_3$ solubility—to promote the formation of insoluble $Cr(OH)_3$. The precipitated $Cr(OH)_3$ is removed from the treated water by a membrane or other filtration process, and the amount of chromium entering the distribution system is lessened. The amount of chromium removal depends on the size of the $Cr(OH)_3$ precipitate, $Cr(OH)_3$ solubility, and the efficiency of the filtration process. Option C is similar to Option B, but a coagulant is added to promote the removal of insoluble $Cr(OH)_3$ through coprecipitation or coagulation followed by filtration. As with Option B, the amount of chromium removal depends on the size of the $Cr(OH)_3$ precipitate, $Cr(OH)_3$ solubility, and the efficiency of the filtration process.

Experimental Protocol

The first step in developing the protocol for the chromium oxidation–reduction (redox) experiments was determining how to collect samples and preserve their speciation for later Cr(VI) analysis. It was decided to optimize the colorimetric method to analyze Cr(VI) immediately after samples were collected. A Spectronic 601 ultraviolet–visible (UV–Vis) spectrometer (Milton Roy, St. Petersburg Fla.) was used with a 10-cm cell instead of the typical 1- or 2.5-cm cell. This longer-path-length cell lowered the detection limit. The method detection limit (MDL) was calculated by multiplying 3.14 by the standard deviation of seven measurements (USEPA 2000), resulting in an MDL of 4 µg/L. This was much better than the 15-µg/L MDL calculated for the Hach DR2010 spectrometer, which uses a 2.5-cm cell.

All experiments were conducted in a constant-temperature room or incubator at 15°C. The experimental protocol included the following steps: A 15-L batch of water was prepared with the appropriate water quality and type of chromium. The water was divided into 500-mL aliquots and poured into high-density polyethylene bottles for either the control or triplicate experiments. The reductant or oxidant to be tested was added to the triplicate samples. The bottles were placed in a tumbler and rotated end-over-end at 30 rpm for the desired reaction time. After the time had elapsed, the bottles were removed from the tumbler, and a 65-mL sample was withdrawn from each bottle. From each 65-mL sample, 30 mL was not filtered—20 mL was used for immediate measurement of Cr(VI) using the colorimetric method on the Spectronic 601, and 10 mL was preserved with 5% nitric acid (HNO_3) for later measurement of total chromium by an inductively coupled plasma–mass spectrometer (ICP–MS). The remaining 35 mL of each sample was filtered through a 0.45-µm nylon syringe filter (Life Science Products, Denver, Colo.)—5 mL was wasted, 20 mL was collected for Cr(VI) measurement, and 10 mL was saved and acid-preserved for the ICP–MS measurement of total chromium. After sampling, the pH of the bottles was adjusted to within 0.1 unit of the desired pH (if required) with 1% HNO_3 or 10 g/L sodium hydroxide (NaOH), and the bottles were returned to the tumbler.

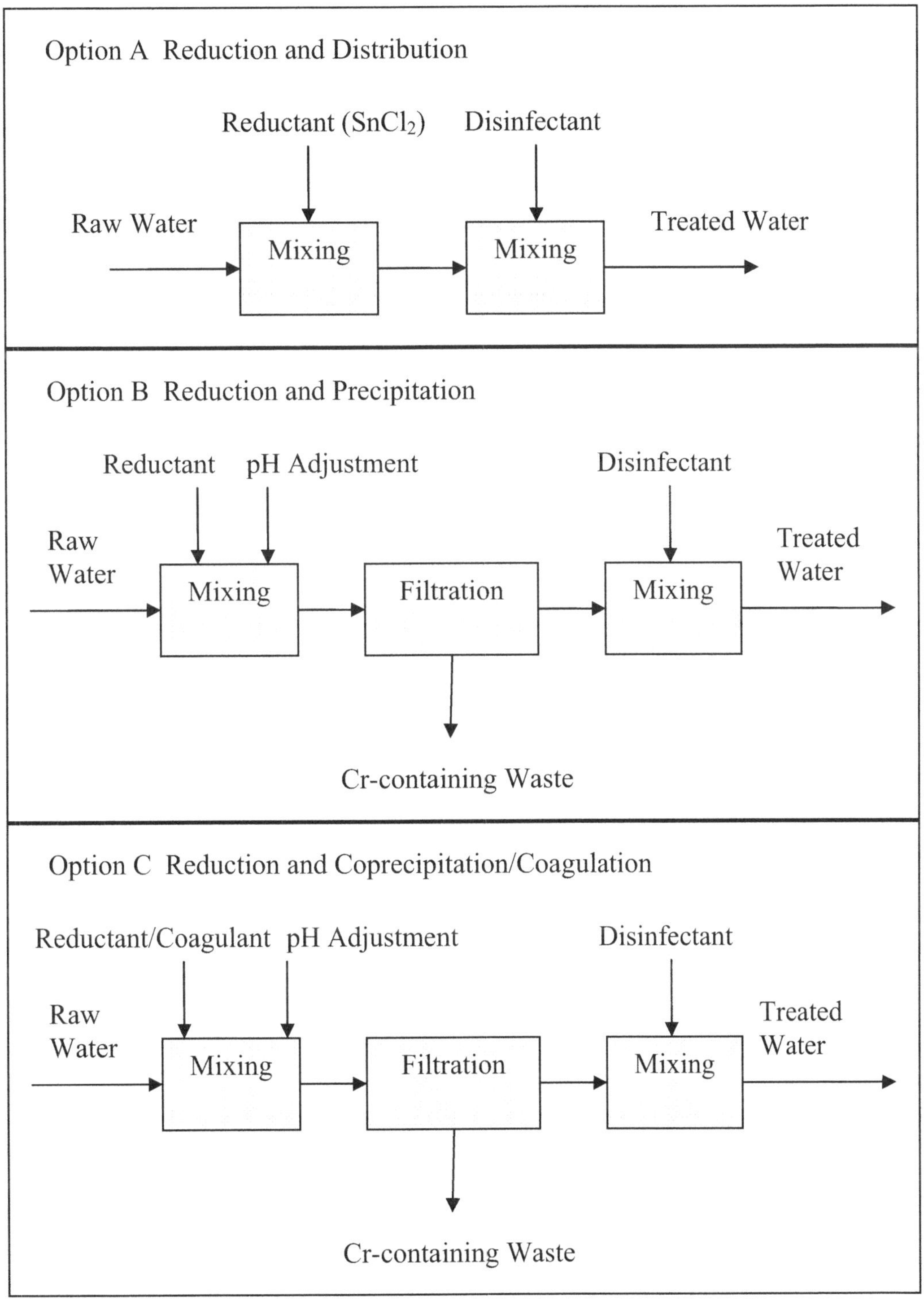

Figure 3.1 Treatment options involving the reduction of Cr(VI) to Cr(III)

There were two exceptions to this protocol. First, in the experiments conducted with deionized (DI) water, only Cr(VI) was measured; total chromium was not measured in these samples. For the reoxidation experiment, a slightly different protocol was used. Reduction of 100 µg/L Cr(VI) to Cr(III) by 300 µg/L Fe^{2+} was carried out in two bottles, each with 2 L of Glendale water. After four days, the solution in each of the two bottles was mixed together in a 5-liter carboy. After the Cr(VI) and total chromium concentrations were measured, the solution was divided into three equal aliquots. The first aliquot was maintained as a control, without the addition of any other chemicals. The second aliquot was dosed with 1 mg/L free chlorine (Cl_2), and the third was dosed with 3 mg/L potassium permanganate ($KMnO_4$), so that any reoxidation of the Cr(III) could be observed. The sampling procedure already outlined was used, except that both 0.45-µm and 0.2-µm filtered samples were collected in anticipation of smaller particles being formed.

At the end of each experiment, the remaining water in each bottle was digested with 5% HNO_3. The goal of the acid digestion was to dissolve any precipitated species that remained in solution and that were not collected during regular sampling, e.g., as a result of particles that were excluded by the pipette tip used for sampling, particles that settled to the bottom of the bottle, or particles that sorbed to the bottle walls or the calcium carbonate ($CaCO_3$) particles present in the synthetic water. Thus, the final concentration of total chromium remaining in the bottle could be determined, and the mass balance of chromium species could be checked.

Conditions Tested

Five water quality matrices were tested. The first was simply 18 mΩ high-purity DI water with 10^{-3} M sodium nitrate ($NaNO_3$) added as a background electrolyte. The second was a laboratory-generated synthetic water prepared to approximate the 75th-percentile concentration of major ions (Table 3.1), based on an analysis of more than 2,000 natural water sources that had at least 5 µg/L of chromium (Chapter 4). Because this synthetic water contained some precipitated $CaCO_3$ at higher pH, a third water quality matrix containing only 1 mg/L $CaCO_3$ was used, so that all of the $CaCO_3$ was soluble. The fourth water quality matrix, the "reducing water," was a modified version of the synthetic water that also contained 5 mg/L TOC (added from a fulvic acid concentrate) and 0.2 mg/L manganese (Mn^{+2}). The fifth water, obtained from Glendale, Calif., is a slightly alkaline water supply with high total dissolved solids (TDS), high alkalinity, high hardness, and low TOC. This water source contains 11 µg/L of naturally occurring Cr(VI), which was supplemented to 100 µg/L total chromium with either Cr(VI) or Cr(III) as appropriate. (The quality parameters of the Glendale water and the median synthetic water are compared in Table 4.5.)

Four oxidants were tested for oxidizing Cr(III) to Cr(VI), and three reductants were tested for reducing Cr(VI) to Cr(III) (Table 3.2).

Chromium Analysis

Samples collected during the experiments were analyzed for Cr(VI) using a Spectronic 201 UV–Vis spectrometer. Two ChromaVer 3 diphenylcarbohydrazide powder pillows (Hach, Loveland, Colo.) were added to 20 mL of sample, and the samples were analyzed at 540 nm after a 5-min reaction time. The instrument was calibrated with six Cr(VI) standards from zero to 120 µg/L, and then a blank and standard were processed as samples. Quality control limits for these analyses included the requirement that the blank must be less than the MDL (4 µg/L) and the standard must be within 10% of the known value.

Table 3.1
Quality of the synthetic water

Constituent	Concentration (mg/L)
Ca	40
Na	86
Mg	30
CO_3	113
SO_4	119
Cl	47
SiO_2	19

Table 3.2
Oxidants and reductants tested

Chemical tested	Dosage	Source
	Oxidants	
Dissolved O_2	Saturated (7 mg/L)	Dissolved O_2
Cl_2	1 mg/L as Cl_2	NaOCl solution
Chloramines	0.5 mg/L as Cl_2, Cl_2:N = 3:1	$(NH_4)_2SO_4$ salt + NaOCl solution
Chloramines—additional tests	0.5, 1, 2, 5, and 10 mg/L as Cl_2	$(NH_4)Cl$ salt + NaOCl solution
MnO_4^-	1, 3, 5, and 10 mg/L as $KMnO_4$	$KMnO_4$ salt
H_2O_2	0.3 w/v % solution	30% H_2O_2 solution
	Reductants	
$SnCl_2$	1.3 mg/L as $SnCl_2$	$SnCl_2.2H_2O$ salt
S^{-2}	1 mg/L as S	$Na_2S.\ 9H_2O$ salt
SO_3^{-2}	1 and 10 mg/L as S	Na_2SO_3 salt

Initial and control samples (i.e., water samples before the reductant or oxidant was added) were analyzed to verify the initial Cr(VI) concentration using a Dionex DX-320 ion chromatograph (IC) with an AD25 post-column UV–Vis detector according to USEPA Method 1636, the Diphenylcarbohydrazide Colorimetric Method (USEPA 1996b). The instrument uses a Dionex Ion Pac NG 1 guard column, a Dionex Ion Pac AS 7 analytical column, a 500-µL sample loop, and a 750-µL reaction coil. The MDL established according to the guidelines (USEPA 2000) was 0.2 µg/L. Samples were preserved with 12 mM soda ash prior to analysis.

For Cr(VI) analysis, the IC was calibrated each day with a blank and at least three standards in a matrix of 12 mM soda ash. The coefficient of determination (r^2) value from the calibration was at least 0.999. Immediately after calibration, the blank and one standard were processed as samples. The blank had to be less than the MDL, and the standard was within 90–110% of the calibrated value. Next, an external reference sample of 10 µg/L Cr(VI) (High Purity Standards, Charleston, S.C.) was run right after the calibration curve to assess ongoing accuracy. The percent recovery on the analysis was within 79–122% of the known value. If these criteria were not met, the run was stopped and the instrument recalibrated. During the run, a blank and one standard were run every ten samples. As noted, the blank had to be less than the MDL, and the standard was within 90–110% of the calibrated value. If these criteria were not met, the samples analyzed after the anomalous blank or standard were rerun.

Samples were analyzed for total chromium using an Agilent 7500c ICP–MS with an octopole reaction system, following USEPA Method 200.8 (USEPA 1999). The octopole system uses helium gas at 6 mL/min as a collision gas to eliminate the $^{40}Ar^{12}C$ mass interference with chromium (^{52}Cr). Germanium (^{74}Ge) was used as the internal standard. The MDL established according to the guidelines (USEPA 2000) was 0.1 μg/L. Samples were preserved with 5% HNO_3 prior to analysis.

The ICP–MS was tuned after each start-up and calibrated with a blank and at least three standards in a 5% HNO_3 matrix. The r^2 value from the calibration was at least 0.999. Immediately after calibration, the blank and one standard were processed as samples. The blank had to be less than 1 μg/L, and the standard was within 90–110% of the calibrated value. Next, an external reference sample of 10 μg/L chromium (High Purity Standards, Charleston, S.C.) was run right after the calibration curve to assess ongoing accuracy. The percent recovery on the analysis was 91–114% of the known value. If these criteria were not met, the run was stopped and the instrument recalibrated. During the run, a blank and one standard were run every ten samples. The blank had to be less than 1 μg/L, and the standard was within 90–110% of the calibrated value. If these criteria were not met, the samples analyzed after the anomalous blank or standard were rerun.

ADSORPTION EXPERIMENTS (CHAPTER 6)

Process Chemistry

Adsorption is a phenomenon in which a contaminant is concentrated at the interface between two phases. In the treatment of Cr(VI), it involves accumulating Cr(VI) on a surface in contact with the contaminated water. Adsorption can be divided into two categories—physical adsorption and chemisorption. Physical adsorption is not site-specific and does not involve the transfer of electrons. Frequently it involves the electrostatic attraction between a charged surface and an ionic species, in this case chromate, or CrO_4^{2-}. As illustrated in Figure 3.2, Cr(VI) anions can potentially be removed by attraction to a positively charged surface.

Chemisorption involves site-specific reactions that exchange electrons. Of the many chemisorption reactions possible, two likely processes were investigated by the project: ligand exchange–chelation and surface reduction and precipitation. Ligand exchange–chelation involves the exchange of one or more ligands (most likely hydroxyl ion groups) for a Cr(VI) ion. If multiple ligands are exchanged with a single ion, the process is called chelation (Figure 3.3).

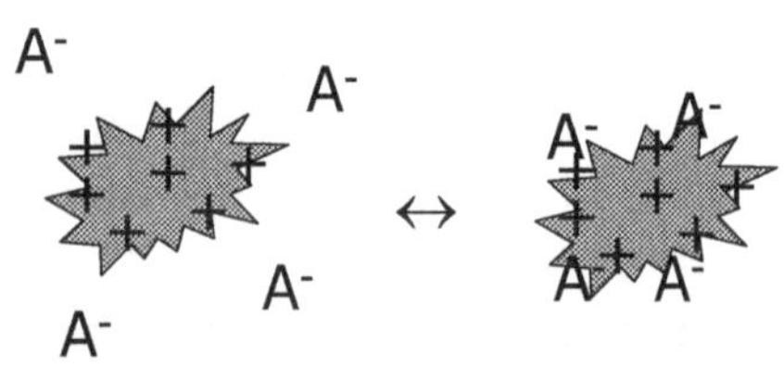

Figure 3.2 Hexavalent chromium treatment by physical adsorption

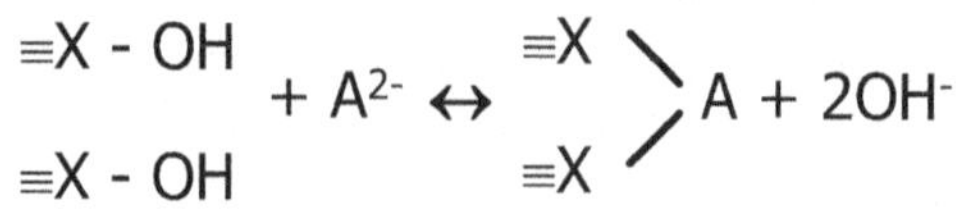

Figure 3.3 Hexavalent chromium treatment by chelation

The hypothesized surface reduction and precipitation process involves two steps. First, Cr(VI) enters into an oxidation–reduction reaction with a reduced metal (likely iron) and is reduced to Cr(III). This in turn forms a precipitate, which is deposited on the surface. Figure 3.4 illustrates the concept of this process.

Process Description

All of the sorption media investigated by the project consisted of column-based technologies designed to operate in a disposable mode—i.e., when the media was exhausted, it would be landfilled and replaced with fresh media. As illustrated in Figure 3.5, adsorption was the simplest Cr(VI) treatment process investigated by the project.

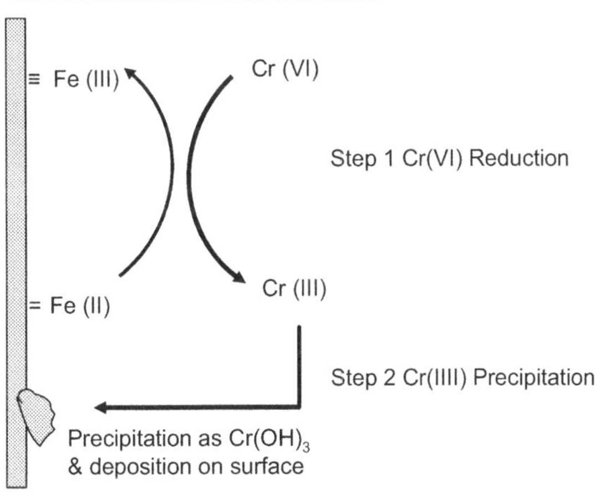

Figure 3.4 Hypothesized mechanism for hexavalent chromium treatment by surface reduction and precipitation

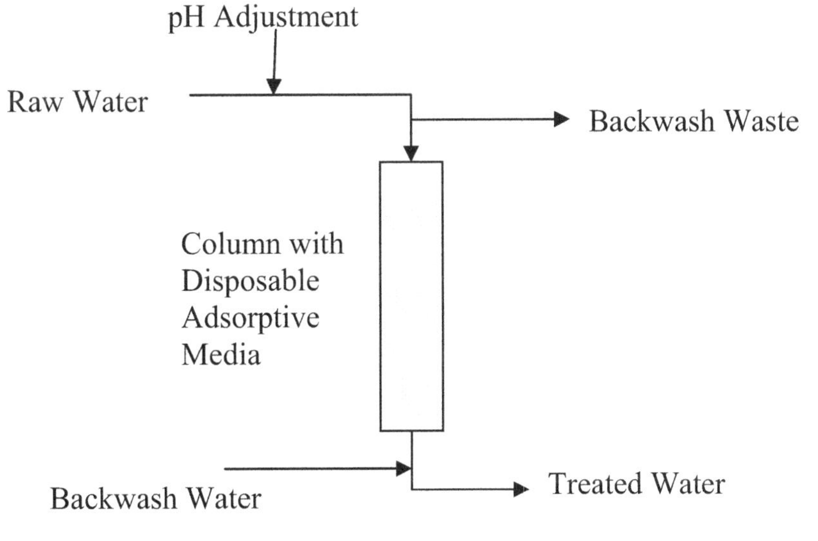

Figure 3.5 Treatment by adsorption or chelation

Batch Experiments

Batch adsorption experiments were used to quantify the Cr(VI) removal capacity of several treatment media. Table 3.3 lists the media evaluated in the batch adsorption experiments. Because most of the media evaluated in this project were developed for arsenic removal and little was known about their capability to remove Cr(VI), these experiments served as a screening mechanism to determine which media removed Cr(VI) from water and which did not.

The batch adsorption experiments consisted of the following individual experiments:

- Initial capacity experiment—This experiment determined if the media removed Cr(VI) from water. If a media did not exhibit the ability to remove Cr(VI) from water, it was not evaluated in subsequent experiments. The initial capacity experiment also provided a starting point for selecting appropriate media dosages for subsequent batch experiments.
- Refined carbonate–buffered capacity experiment—The refined carbonate–buffered experiment was developed to further refine the Cr(VI) removal capacity of media that had demonstrated good performance in the initial capacity experiment.
- pH sensitivity experiment—This experiment examined the efficiency of Cr(VI) removal across a pH range of 5 to 9.
- Dual-solute experiments—This series of experiments examined changes in Cr(VI) removal efficiency in the presence of competing solutes. It also evaluated the media performance at removing competing solutes.

Table 3.3
Treatment media evaluated in batch adsorption experiments

Media Identification	Media description	Media type
Media A	Granular ferric hydroxide	Sorption
Media B	Granular ferric oxide	Sorption
Media C	Sulfur-modified iron	Surface reduction and precipitation
Media D	Activated alumina	Sorption
Media E	Iron-impregnated activated alumina	Sorption
Media F	Iron-impregnated zeolite	Surface reduction and precipitation
Media G	Iron oxide–coated diatomite	Sorption
Media H	Bauxite clay	Sorption
Media J	Metal-binding ligand	Chelation

All of the batch adsorption experiments followed the same experimental protocol. The protocol began with a 500-mL plastic bottle filled with stock water containing 100 μg/L Cr(VI). Additional characteristics of the stock water were varied in each of the individual experiments. A known dose of adsorptive media was added to the bottle, and the bottle was placed on a rotating bottle tumbler that rotated at 30 rpm to ensure the media were suspended sufficiently in the stock water. At 30 min, 2 hr, and in some cases 24 hr, the bottle was removed from the tumbler for sample analysis. First, pH was measured. Then a small-volume water sample was withdrawn from the bottle with a syringe and filtered through a 0.45-μm syringe filter. The Cr(VI) in the filtered sample was measured with a Hach DR4000 spectrophotometer by means of the Hach 8023 method, a colorimetric method using diphenylcarbazide. Additional filtered sample volume was held for further sample analysis such as total chromium analysis by ICP–MS, low-level Cr(VI) analysis by IC, or additional solute analysis by a variety of analytical methods. After sample analysis, the bottle was returned to the rotating bottle tumbler.

Figure 3.6 describes the protocol used for the batch adsorption experiments. Specific details of individual experiments are described in the following sections.

Initial Capacity Experiment

The initial capacity experiment determined if the adsorptive media removed Cr(VI) from water. Media that did not exhibit the ability to remove Cr(VI) from water were not evaluated in subsequent experiments. The initial capacity experiment also served as a starting point for selecting appropriate media dosages for subsequent batch experiments. Media dosages selected for the initial capacity experiment were 0.5, 1, 2, 4, and 8 g/L. Stock water for this experiment consisted of Milli-Q water with 100 μg/L Cr(VI). Sodium chromate was used as the source of Cr(VI). The stock water for this experiment was not buffered or pH-adjusted. Concentrations of Cr(VI) were sampled at 30 min, 2 hr, and 24 hr.

Refined Carbonate–Buffered Capacity Experiment

To build on the information developed in the initial capacity experiment, a refined carbonate–buffered capacity experiment was developed to maintain stable pH. Stock water for this experiment consisted of Milli-Q water with 100 μg/L Cr(VI) buffered with 1 mM bicarbonate. Sodium bicarbonate was used to create the bicarbonate-buffered solutions. Media dosages were lowered to 0.25, 0.5, 1, 2, and 4 g/L to increase the likelihood of measurable Cr(VI) concentrations at the 2- and 24-hr sampling times.

pH Sensitivity Experiment

The pH sensitivity experiment examined Cr(VI) removal efficiency at initial pH conditions of 5, 7, and 9. Stock water for this experiment consisted of Milli-Q water buffered with 1 mM bicarbonate, adjusted to the selected initial pH condition, and spiked with 100 μg/L Cr(VI). The dosage for all treatment media was 1 g/L. Cr(VI) was sampled at 30 min and 2 hr.

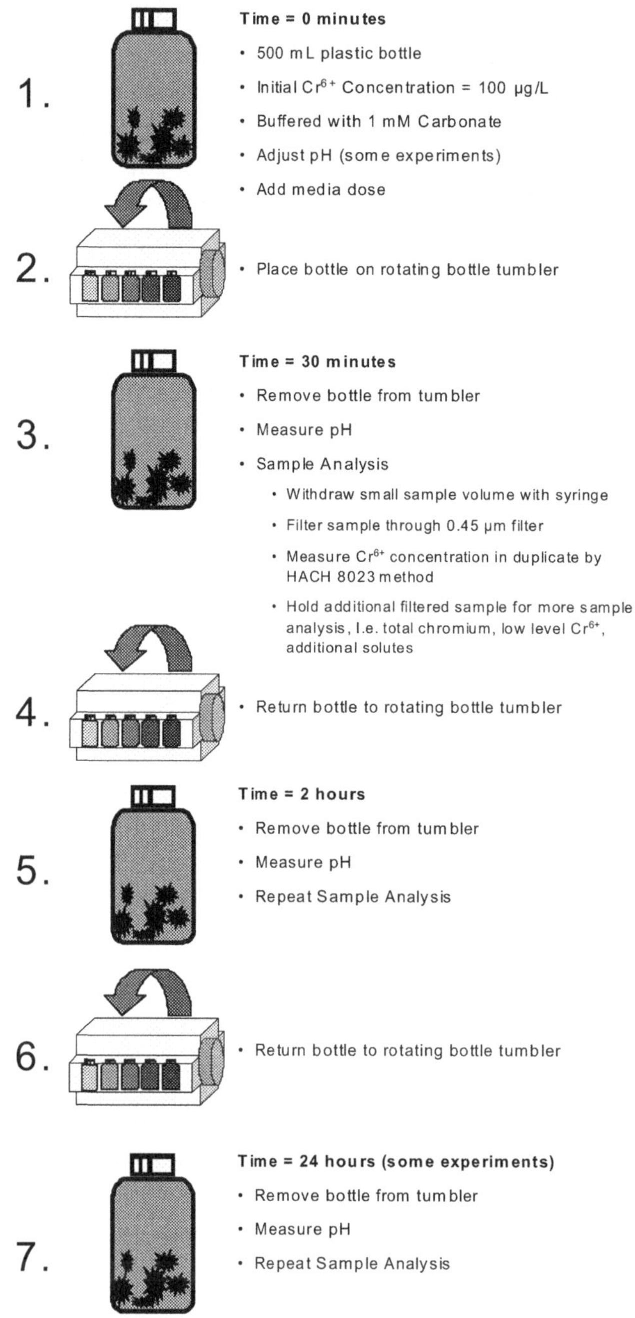

Figure 3.6 Protocol for batch adsorption experiments

Dual Solute Experiments

A series of dual solute experiments examined the efficiency of Cr(VI) removal in the presence of competing solutes at three concentrations. These experiments also evaluated media performance at removing the competing solutes. Previous experience with the adsorptive media used to remove arsenic indicated decreased performance in the presence of various co-occurring solutes. It was also important to learn whether the adsorptive media were able to remove other contaminants of concern. Table 3.4 describes the two categories of dual solutes evaluated in this series of experiments: common co-occurring solutes and solutes for which co-removal would be advantageous. The corresponding low, medium, and high initial concentrations of each solute are also presented.

The initial concentrations shown in Table 3.4 were chosen to represent a range of conditions expected in drinking water sources containing Cr(VI).

Stock water for these experiments consisted of Milli-Q water buffered with 1 mM bicarbonate, adjusted to pH 7, and spiked with 100 μg/L Cr(VI). Competing solutes were dosed at low, medium, and high concentrations. The dosage for all media was 1 g/L, except for media C, which had a dosage of 0.1 g/L. Cr(VI) and the competing solutes were sampled at 30 min and 2 hr.

Column Experiments

Sulfur-Modified Iron

The sulfur-modified iron treatment media was evaluated in a column operation because it demonstrated excellent performance at bench-scale. The experimental program to evaluate Cr(VI) removal with the sulfur-modified treatment media in a column began with a preliminary experiment to determine media capacity. This was followed by another experiment using two parallel columns that examined pH effects. Batch isotherm experiments were also conducted to elucidate results from the column experiments.

All of the column experiments with sulfur-modified iron used the same column design, which was based on previous experimental work with the sulfur-modified iron treatment media conducted by Cindy Schreier at Prima Environmental. The Prima Environmental column design was developed to prove the feasibility of low-level Cr(VI) removal and had the following characteristics:

- Upflow configuration
- 1-in. syringe columns
- Empty bed contact times (EBCTs): 3, 5, and 7 min
- Flow rate: 6.6 mL/min
- Influent Cr(VI) concentration: 100 μg/L

The column design for this project, based on the Prima Environmental column design, was developed to determine media capacity in a short time frame and had the following characteristics:

- Upflow configuration
- ½-in. PVC columns
- EBCT: 2 min
- Flow rate: 10 mL/min
- Influent Cr(VI) concentration: 1.00 to 2.67 mg/L

Five column tests were conducted with sulfur-modified iron treatment media. Table 3.5 describes the differences between the column tests.

Table 3.4
Dual solute concentrations evaluated in dual solute experiments

Constituent	Initial concentration			Unit
	Low	Medium	High	
Common co-occurring solutes				
Chloride	10	50	100	mg/L
Sulfate	10	25	50	mg/L
Bicarbonate	10	100	250	mg/L
Silicate	5	10	25	mg/L
TOC	0.5	1	5	mg/L
Phosphate	0.05	0.2	0.4	mg/L
Co-removal solutes				
Arsenic	10	25	50	μg/L
Nitrate	2.5	5	10	mg/L
Fluoride	0.5	2.5	5	mg/L
Perchlorate	10	50	100	μg/L

Table 3.5
Column experiments on sulfur-modified iron treatment media

Experiment	Column	Influent Cr(VI) concentration (mg/L)	Influent pH
Initial capacity column experiment	Initial	2.67	8.3
pH effect column experiment	A	1.00	8.3
	B	1.00	7.0
Confirmatory pH effect column experiment	C	2.67	8.3
	D	2.67	7.0

The following sections describe specific details of each column experiment using the sulfur-modified iron treatment media.

Initial Capacity Column Experiment

The initial capacity column experiment was designed to determine media capacity for Cr(VI) in a short time frame. To achieve this objective, the column was operated with a shorter EBCT, higher flow rate, and higher influent Cr(VI) concentration than used with the Prima Environmental column design. Stock water for this experiment consisted of Milli-Q water with 2.67 mg/L Cr(VI) and buffered with 1 mM bicarbonate. Sodium bicarbonate was used to add bicarbonate ions into solution. Initial pH was not adjusted in the buffered solution and remained at 8.3.

pH Effect Column Experiment

After an initial estimate of media capacity had been established with the initial capacity column experiment, two columns (A and B) with the same design used in the initial capacity experiment were operated in parallel to determine pH effects. Stock water for this experiment consisted of Milli-Q water with 1.00 mg/L Cr(VI) and buffered with 1 mM bicarbonate. Sodium bicarbonate was used as the bicarbonate source. Initial pH was adjusted in the buffered solution to 8.3 and 7.0 for columns A and B, respectively.

Confirmatory pH Effect Column Experiment

An additional column experiment on pH effect was conducted at the higher Cr(VI) concentration to validate the results observed in the previous column experiment on pH effect. For this experiment, two columns with the same design used in the initial capacity experiment were operated in parallel to determine pH effects. Stock water for this experiment consisted of Milli-Q water with 2.67 mg/L Cr(VI) and buffered with 1 mM bicarbonate. Sodium bicarbonate was used to add bicarbonate ions into solution. Initial pH was adjusted in the buffered solution to 8.3 and 7.0 for columns C and D, respectively. Column C also served as a duplicate experiment for the initial capacity experiment.

Batch Isotherm Experiments

Batch isotherm experiments were conducted to help clarify the conflicting mass loading or capacity estimates developed from the column tests on sulfur-modified iron treatment media. These batch experiments followed the same protocol as the batch adsorption experiment on pH sensitivity but at the higher initial Cr(VI) concentrations used during the column experiments. Stock water for this experiment consisted of Milli-Q water with 2.67 mg/L Cr(VI) and buffered with 1 mM bicarbonate. Sodium bicarbonate was used as the bicarbonate source. Initial pH was adjusted in the buffered solution to 8.3 and 7.0. Media dosages were 0.1, 0.2, 0.3, 0.4, and 0.5 g/L. Cr(VI) was sampled at 30 min, 2 hr, and 24 hr.

Granular Ferric Hydroxide Experiment

The granular ferric hydroxide treatment media was evaluated in a column operation to confirm the results demonstrated in the bench-scale batch adsorption experiments. The column design for the granular ferric hydroxide experiment had the following characteristics:

- Two downflow columns in series
- Column diameter: 1 in. (2.54 cm)
- EBCT: 2.5 min for each column, for a 5.0-min total EBCT
- Superficial velocity: 2 gpm/sq ft (5 m/hr)
- Feed water: Boulder tap water spiked with 100 μg/L Cr(VI) and 100 μg/L arsenic
    - pH: ~7.7
    - TOC: ~0.5 mg/L
    - Sulfate: <30 mg/L
    - Bicarbonate: <50 mg/L as $CaCO_3$
    - Chlorine residual: 1.0 mg/L

Exhausted Media Leaching and Digestion Experiment

After the initial and confirmatory column experiments on pH effect, the exhausted media used in the column experiments were analyzed for their leaching potential and digested completely to determine chromium mass loading on the media. Columns A, B, C, and D were capped and shipped to Utah State University, where they were cut into quarter-sections labeled according to the upflow column operation: quarter-section 1 at the bottom and quarter-section 4 at the top. Media from each quarter-section underwent total digestion, and column composite samples were used for leachate testing. Total digestion of the media samples was accomplished using hydroxylamine digestion. The leachate testing protocol followed two approaches: (1) the USEPA Toxic Characteristic Leaching Protocol (TCLP) (Method 1311) and (2) the California Waste Extraction Test (WET). Because of the small amount of media available within the columns, the samples sizes used for the TCLP and the WET were reduced as described in Table 3.6. The reagent volumes were scaled down proportionally.

The leachate resulting from the total digestions and the TCLP and WET were analyzed by ICP–MS for total chromium at dilutions of 10:1, 100:1, and 1,000:1.

ANION EXCHANGE EXPERIMENTS (CHAPTER 6)

Treatment Chemistry

Ion exchange is a process involving the exchange of ions from one phase to another. In typical water treatment applications, the process involves the exchange of ionic contaminants in the liquid phase (water) for harmless ions in the solid phase (resin). The solid phase is a synthetic resin backbone to which charged sites (charged functional groups) are attached. Initially, these charged functional groups are saturated with noncontaminant ions such as chloride. As water passes through the resin, the noncontaminant ions are exchanged with the contaminant ions, which bind more strongly with the resin. When all the exchange sites have been filled with the contaminant ion, the ion exchange resin bed is exhausted. The contaminant ion can be displaced from the resin and the resin bed regenerated by exposing the resin to high concentrations of the noncontaminant ion. After regeneration, the resin can be reused to remove the contaminant.

The generally accepted ion exchange treatment process for Cr(VI) consists of exchanging the chloride (Cl^-) anions for chromate ($HCrO_4^-$) anions through the use of strong base anion exchange resins (SBA), as shown in Equation 3.3.

$$\equiv X^+ - Cl^- + HCrO_4^- \leftrightarrow \equiv X^+ - HCrO_4^- + Cl^- \qquad (3.3)$$

SBA resins are so named because their behavior resembles that of a strong base resin, and their functional groups remain ionized over a wide range of pH values. SBA resins are classified into two groups, Type I and Type II. Type I resins consist of quaternary amine functional groups; Type II resins replace a methyl group with an ethanol group. Thus, Type II resins are slightly less basic than Type I resins. Type I resins typically have higher affinity for contaminants such as carbonate or silica, whereas Type II resins are easier to regenerate.

Table 3.6
Sulfur-modified iron media sample sizes for the TCLP and WET

Column	Sample size for the TCLP (grams)	Sample size for the WET (grams)
A	20	20
B	50	20
C	20	20
D	20	20

Although not investigated in this bench-scale work, weak base anion exchange resins are theoretically capable of treating Cr(VI). However, the highest pH at which their functional groups remain charged is approximately pH 6. Therefore, treatment should occur at pH 6 or lower to take advantage of the ion exchange capacity as it is currently understood. The greater capacity of weak base anion exchange resins gives them a potential advantage over SBA resins for chromium control.

Process Description

Conventional anion exchange treatment is implemented in a column configuration. As illustrated in Figure 3.7, chromium is removed by continuously passing untreated water through a column packed with resin. When the column exchange capacity is reached, regeneration is initiated by passing a high concentration (approximately 1N) chloride solution through the column. Typically, two equivalents of chloride per equivalent of resin are required to complete regeneration. Four to five bed volumes of brine and rinse water are created for each regeneration.

The performance of the resin can be influenced by the presence of competing anions. However, chromium is generally preferred over other prevalent anions such as sulfate. The use of anion exchange for treating chromate is likely to be effective even in water supplies with elevated concentrations of sulfate.

Column Experiments

Four commercially available SBA exchange resins were evaluated for their ability to remove Cr(VI) from a synthetic water supply. Detailed properties of the resins are included in Table 3.7. The experiments were conducted with 0.5-in.-diameter mini-columns operated in a downflow mode with an EBCT of 2.5 min and a hydraulic loading rate of 1.0 gpm/sq ft.

The columns were supplied with synthetic water consisting of DI water spiked with 1,000 μg/L sodium chromate, 200 mg/L sodium bicarbonate, and 200 mg/L sodium sulfate. The columns were operated continuously to exhaustion.

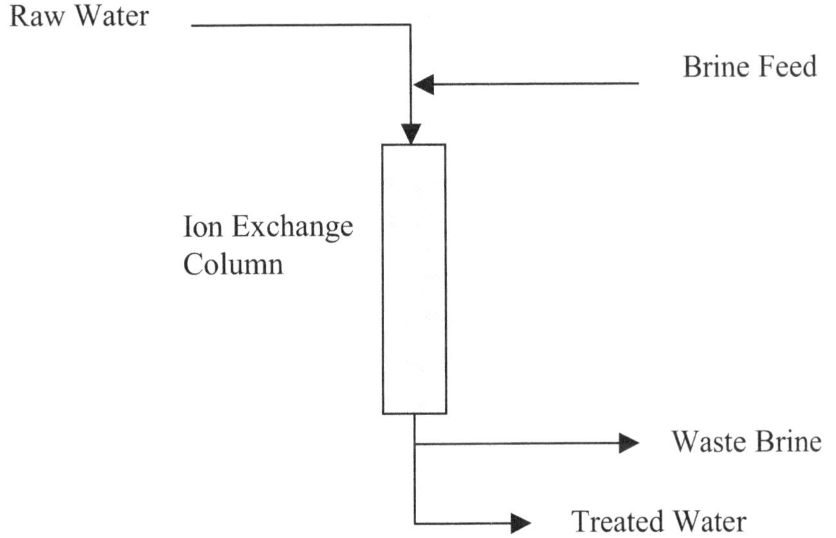

Figure 3.7 Conventional ion exchange process

Table 3.7
Properties of ion exchange resins used in column experiments on conventional anion exchange

Properties	Marathon MSA	Amberlite IRA410	A-600	A-300
Manufacturer	DOW	Rohm & Haas	Purolite	Purolite
Description	SBA Type I	SBA Type II	SBA Type I	SBA Type II
Matrix	Polystyrene divinylbenzene copolymer	Styrene divinylbenzene copolymer	Polystyrene divinylbenzene copolymer	Polystyrene divinylbenzene copolymer
Type	Macroporous	Gel	Gel	Gel
Functional group	Quaternary ammonium (?)	Quaternary amine	Quaternary ammonium	Quaternary amine
Ionic form	Chloride	Chloride	Chloride	Chloride
Total exchange capacity (meq/mL)	1.1	1.25	1.4	1.45–1.6
Screen grading	-	16–50 mesh	16–50 mesh	16–50 mesh
Mean particle size (μm)	640 ± 50	-	-	-

Reactor Experiments (MIEX® resin)

The project also investigated an innovative treatment process using a continuously mixed resin treatment process (Figure 3.8). In this process, fresh or regenerated anion exchange resin is continuously fed into a mixed contactor, exposing the resin to Cr(VI)-contaminated water. After a hydraulic retention time of 20 min, the resin–treated water mixture is passed through a settling unit, where the resin is separated from the treated water and returned to the contactor. Ten percent of the recycled resin is diverted from the recycle stream and regenerated. After regeneration, the resin is returned to the contactor.

The magnetic ion exchange resin treatment media was evaluated in a small-scale pilot unit to demonstrate performance under continuous operation.

The small-scale pilot unit (Figure 3.9) was operated under the following conditions:

- Flow: 2 L/min
- Water source: Boulder tap water spiked with 100 μg/L Cr(VI)
  - pH: 7.7
  - Sulfate: <30 mg/L
  - Bicarbonate: <50 mg/L as $CaCO_3$
  - Chlorine residual: 1.0 mg/L
- Resin dosage: 20 mL/L
- Contact time: 20 min
- Regeneration rate: 10%

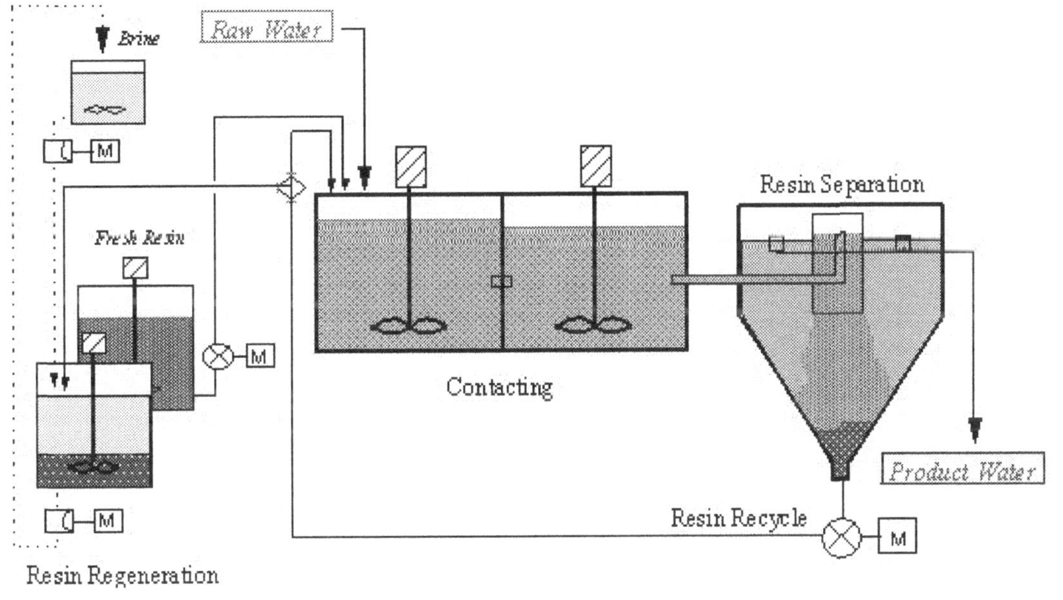

Figure 3.8 Continuously mixed ion exchange process (MIEX®)

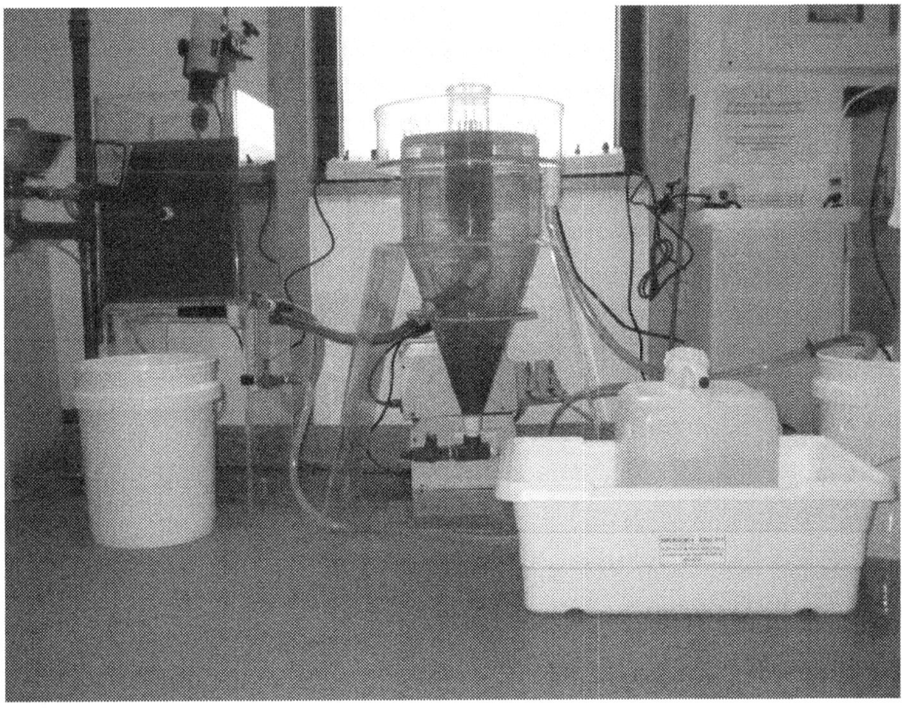

Figure 3.9 Small-scale MIEX® pilot unit

MEMBRANE TECHNOLOGY EXPERIMENTS (CHAPTER 7)

Treatment Chemistry

Reverse osmosis and nanofiltration (NF) are pressure-driven separation processes in which a membrane serves as a barrier that retains dissolved solutes but allows water to pass through. RO and NF use two distinct mechanisms for Cr(VI) removal: (1) electrostatic repulsion, which is the rejection of ions by the membrane owing to the charge of the species, and (2) physical straining, which is the rejection of molecules that are larger than the membrane pore size. Because the predominant removal mechanism in NF is electrostatic repulsion, the anionic character of Cr(VI) is the fundamental reason that NF could be effective for Cr(VI) control. Chromium oxidation state is less important for RO performance because the pores are so small that chromium molecules are hindered from passing through the membrane by their size alone. RO and NF membranes are also effective at removing co-contaminants such as arsenic, perchlorate, and nitrate.

Process Description

RO and NF treatment consists of three processes: pretreatment, membrane treatment, and post-treatment (Figure 3.10). Rigorous pretreatment is usually required to prevent fouling of the membrane. The pretreatment step includes the addition of acid or antiscalants to inhibit the formation of inorganic precipitates such as calcium carbonate, calcium sulfate, strontium sulfate, or barium sulfate. If the untreated water contains moderate or high concentrations of suspended solids or particulate matter, pretreatment by conventional methods or microfiltration is required.

For the treatment step, RO and NF membranes are packaged in spiral-wound elements that are 40 in. long and either 4 or 8 in. in diameter. Several elements are connected in series within pressure vessels, and these pressure vessels are staged in an array configuration (Figure 3.11). Water that has passed through the membrane (the permeate) is sent to the distribution system. Water that has been retained by the membrane (the retentate) is sent to the next stage of the array, where it is treated further. This process continues until 80–85% of the water supplied to the unit has passed through the membrane.

Following membrane treatment, a post-treatment step is required. Water supplies treated by RO and NF are typically low in pH and unstable with respect to corrosion potential, so adjustments are made to increase the pH and alkalinity of the treated water. Then the water is disinfected before entering the distribution system.

Membranes

Six membranes—two RO, three NF, and one tight ultrafiltration (UF)—were tested in this research. The RO membranes were selected to reflect a range of operating pressures, and the NF and UF membranes were selected to reflect differences in molecular mass/weight cutoff (MMCO/MWCO) and membrane surface charge. Each membrane was identified and selected according to information from the manufacturer's catalogue. Specimens of each membrane were independently characterized according to pure water permeability (L/m^2-day-kPa), polyethylene glycol rejection (MMCO/MWCO), zeta potential (surface charge), contact angle (hydrophilicity/hydrophobicity), and attenuated total reflection–Fourier transform infrared (ATR–FTIR) analysis (functional groups). Table 3.8 describes the physical–chemical properties of the membranes evaluated.

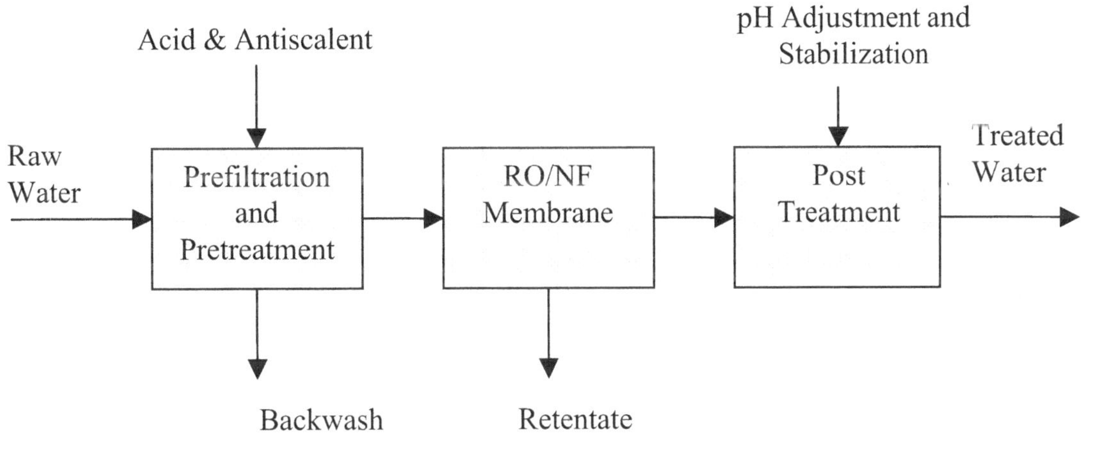

Figure 3.10 Reverse osmosis–nanofiltration process

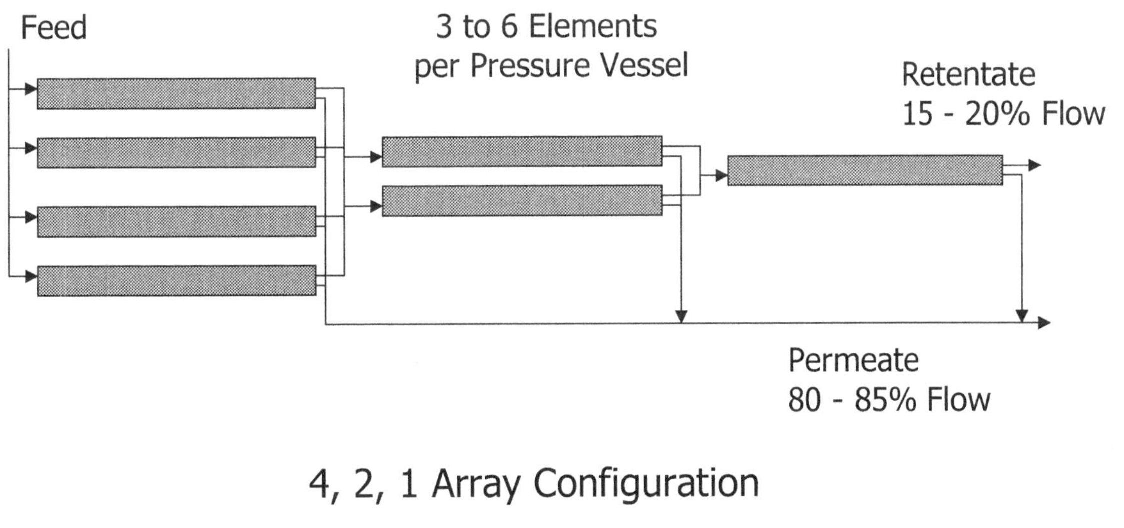

Figure 3.11 Membrane array

Table 3.8
Characteristics of the membranes tested

Membrane	Manufacturer	Material	Molecular mass cutoff*	Contact angle (°)†	PWP (L/m^2-day-kPa)‡	Zeta potential (mV)§
LFC-1 (RO)	Hydranautics	Polyamide TFC	n/a	70.3	0.75	–4.5
TFC-ULP (RO)	Fluid System	Polyamide TFC	n/a	48	1.56	–28.3
TFC-S (NF)	Fluid System	Polyamide TFC	200	57	1.99	–23.5
ESNA (NF)	Hydranautics	Polyamide TFC	200	57	1.05	–11.1
MX07 (NF)	Osmosis	Polyamide TFC	400	45.3	0.47	–36.8
GM (UF)	Desal	Proprietary	8,000	46	2.81	–32.2

* Manufacturer's data
† Average value (3 measurements)
‡ Pure water permeability at room temperature (~20°C)
§ Measured at pH 8 and conductivity 30 millisiemens per meter (mS/m) with potassium chloride
n/a = not available

Water Supplies

Several 0.45-μm prefiltered water supplies were used to conduct bench-scale, cross-flow flat-sheet filtrations:

- Colorado River water (CRW) from the Metropolitan Water District (MWD) of Southern California,
- Groundwater from LADWP,
- Groundwater from Glendale, Calif., and
- Various synthetic waters.

Synthetic water supplies were prepared with Milli-Q (inorganic and organic carbon–free water, which was deionized with two proprietary cation exchange beds and an anion exchange bed followed by filtration with a 0.2-μm filter. Among the three natural water sources, the LADWP and Glendale water supplies were contaminated with 120 μg/L and 70 μg/L of Cr(VI) as chromate, comprising about 75~80% and 45~50% of the total chromium, respectively, whereas the CRW contained 4–9 μg/L of perchlorate. In addition to Cr(VI), two other toxic anions, arsenate and perchlorate, were added to the test solutions.

The target anions (present at total concentrations of 100 μg/L of chromate, arsenate, and perchlorate after spiking) were introduced to the membranes as a single component, in binary mixtures with other electrolytes (KCl, K_2SO_4, and $CaCl_2$), and at varying pH values (4, 6, 8, and 10) and solution conductivities (ionic strength) (30, 60, and 115 millisiemens per meter [mS/m]). Either potassium hydroxide (KOH) or hydrochloric acid (HCl) was used to adjust pH. Table 3.9 summarizes the water quality parameters of the water supplies tested in this study.

Table 3.10 shows the concentrations of the electrolytes used to adjust conductivity in the membrane experiments. Cations and anions in the water supplies were measured to predict how they would influence rejection/transport of the target anions. As shown in Table 3.11, the natural water supplies exhibited various concentrations of both monovalent and divalent ions.

Table 3.9
Quality parameters of the synthetic and natural water supplies tested

Source water	pH	Conductivity, mS/m	TDS, mg/L	DOC, mg/L	UVA_{254}, cm^{-1}
Synthetic water	4, 6, 8, and 10	30, 60, and 115[d]	NA	NA	NA
CRW[a]	7.93	74.9	359	2.44	0.045
LADWP[b]	7.96	75.5	365	0.62	0.015
Glendale[c]	7.27	77.7	375	0.47	0.01

[a] Fall season, 2001 (CRW)
[b] Summer season, 2002 (LADWP)
[c] Fall season, 2002 (Glendale)
[d] Conductivity adjusted with KCl, K_2SO_4, and $CaCl_2$
NA = not available

Table 3.10
Amounts of salts used to adjust conductivity

	Salt Concentrations in mg/L to Achieve Target Conductivities		
Salt	30 mS/m	60 mS/m	115 mS/m
KCl	149.1	298.2	574.0
K2SO4	209.1	418.2	801.6
CaCl2	155.4	310.8	599.3

Table 3.11
Ion concentrations in feed waters

Ions, mg/L	CRW[a]	LADWP[a]	Glendale[a]
Silica (SiO_2)	3.2	10.5	17.1
Barium (Ba^{2+})	0.007	0.12	0.05
Calcium (Ca^{2+})	55.1	91.7	89.9
Magnesium (Mg^{2+})	21.4	20.1	29.2
Sodium (Na^{+})	47	26.1	40.6
Potassium (K^{+})	3.4	4.3	2.4
Sulfate (SO_4^{2-})	190	68.7	105.7
Nitrate (NO_3^{-})	0.28	58.2	37.4
Chloride (Cl^{-})	60.5	34.2	64.3
Chromate (CrO_4^{2-})	0[b] /100[b]	0.12[b] /0.12[b]	0.07[b] /0.1[b]
Arsenate ($HAsO_4^{2-}$)	0[b] /0.1[c]	0[b] /0.1[c]	0[b] /0.1[c]
Perchlorate (ClO_4^{-})	0.004[b] /0.1[c]	0[b] /0.1[c]	0[b] /0.1[c]

[a] All feed waters are from the same season as those in Table 3.9.
[b] Ambient concentrations of chromate, arsenate, and perchlorate in feed waters
[c] Total concentrations of chromate, arsenate, and perchlorate after spiking

Membrane Testing Units

Stirred Cell Unit

Figure 3.12 is a schematic of a dead-end stirred cell unit (Millipore). The stirred cell was used to conduct polyethylene glycol tests to determine the real MMCO/MWCO of the membranes.

Continuous Cross-Flow Flat-Sheet Filtration Membrane Testing Unit

Figure 3.13 is a schematic of the experimental system used in flat-sheet tests at constant or variable pressures or both. An Osmonics SEPA flat-sheet test cell was used for the cross-flow experiments. This cell permits testing of a single 9.5-cm × 14.6-cm membrane coupon under tangential feed flow conditions with a channel height of 0.142 cm. The module embodies an active filtration area of 139 cm^2.

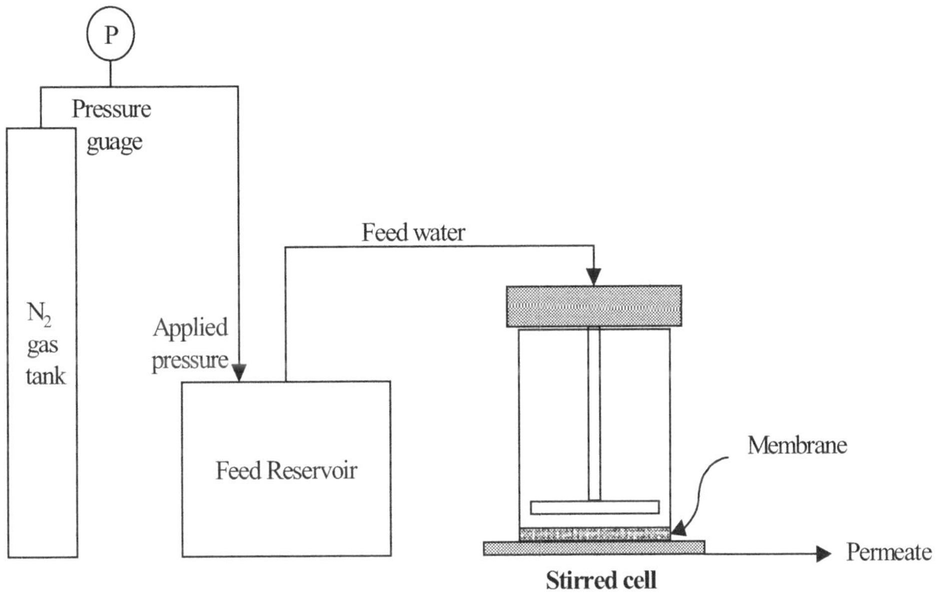

Figure 3.12 Schematic of stirred-cell filtration unit

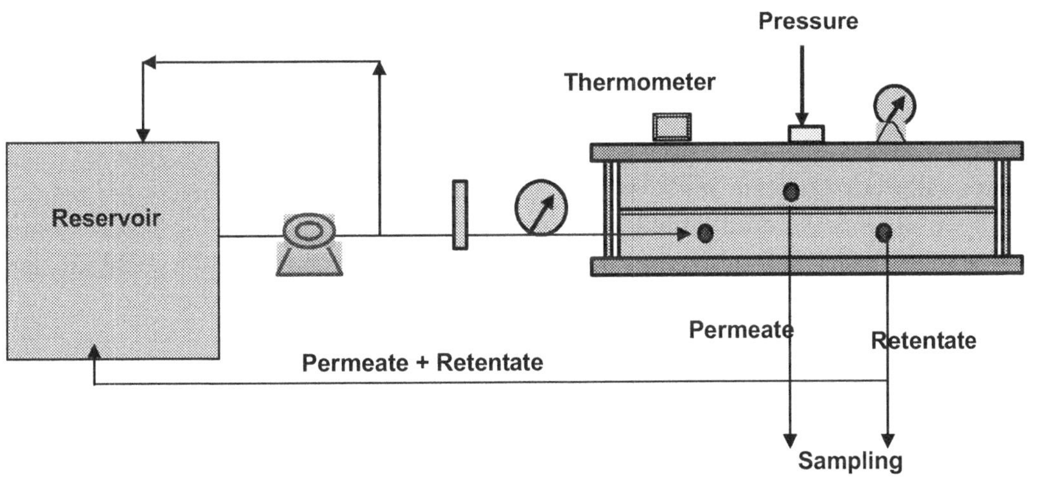

Figure 3.13 Schematic of experimental flat-sheet test cell

Each virgin membrane tested was prefiltered with a pure water volume of 8 L without recirculation through the flat-sheet unit (a feed flow rate of ~200 mL/min and a permeate flow rate of ~5 mL/min) to remove any preservative materials that might have become coated on or in the membrane surface or pores. Then the new membrane was soaked in pure water for at least 24 hr. During this period, the pure water was replaced every 2 or 3 hr with another new volume of pure water for membrane stabilization and the new membrane was filtered again overnight with pure water. The pure water permeate flow was monitored for several hours just before each test until constant pure water flux was obtained.

The feed solution was drawn from a 4-L reservoir and supplied to the flat-sheet module. The retentate and permeate streams were returned (or discarded) to the reservoir after passing through the test cell. The temperature of the feed water was maintained at 20 ± 1°C by immersing the reservoir in a temperature-controlled water bath. To assure thermal equilibration, the source water was kept at room temperature for 24 hr prior to the membrane tests. The retentate flow, permeate flow, and temperature were monitored over time, and the feed flow was calculated from the measured retentate and permeate flows. Observed solute rejection, R_{obs}, was calculated by using Equation 3.4

$$R_{obs}(\%) = \frac{C_f - C_p}{C_f} \times 100(\%) \tag{3.4}$$

in which C_f is the solute concentration in the feed and C_p is the solute concentration in the permeate.

Membrane Oxidative Damage

These experiments, conducted to assess the effects of long-term Cr(VI) exposure to polymeric membranes, were performed with synthetic waters spiked with 0.1, 100, and 10,000 mg/L of Cr(VI) and adjusted to various pH levels (4, 6, 8, and 10) with HCl or NaOH. The RO, NF, and UF membranes were soaked at each of the different conditions, flux and NaCl rejection were monitored, and the membrane surfaces were characterized by FTIR analysis and scanning electron microscopy as a function of CT (concentration × time) for 12 months. Membrane flux and NaCl rejection were compared before and after exposure at various Cr(VI) concentrations and pH ranges.

As shown in Figure 3.14, this study was conducted under 12 different conditions. Cr(VI) is present in water as diprotic chromic acids species, with monovalent $HCrO_4^-$ below pH 6.5 and divalent CrO_4^{2-} above pH 6.5. An additional Cr(VI) species, dichromate ($Cr_2O_7^{2-}$), predominates at concentrations greater than 1,000 mg/L, but no drinking water source is likely to contain such a high concentration of Cr(VI).

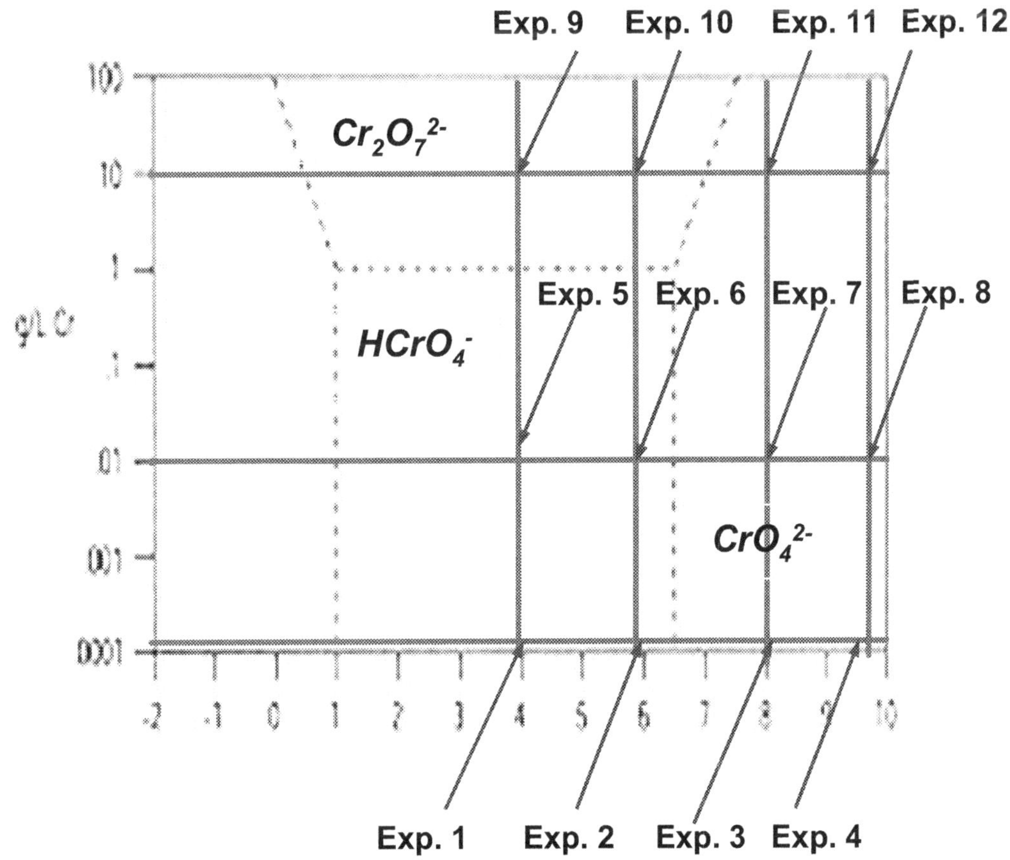

Figure 3.14 Distribution of Cr(VI) species and experimental conditions as functions of pH and concentrations

REDUCTION AND PRECIPITATION EXPERIMENTS (CHAPTER 8)

The study compared the performance of zerovalent and divalent iron—Fe(0) and Fe(II)—as reductants, evaluated reduction kinetics, and examined the formation of Cr(III) precipitates and coprecipitates. The process variables investigated included pH, iron dosage, and the presence of co-occurring contaminants including sulfate, phosphate, arsenate, and silica.

The experimental work initially consisted of four phases. Phases A and B compared the performance of Fe(0) and Fe(II) as reductants and examined how pH, reductant dosage, and the presence of co-occurring parameters influenced the reduction process. Phase C involved simulations of the reduction and precipitation process for chromium removal, using a membrane filter to remove precipitated or coprecipitated Cr(III) after Cr(VI) reduction. Phase D studied the kinetics of Cr(VI) reduction in a natural water supply using Fe(II). It also investigated the chromium removal process in more detail, including ways to optimize the process. Table 3.9 summarizes the initial experimental plan.

Phase E—Cr(VI) Reduction Experiments in Glendale Water

Phase E was added to the experimental plan to evaluate additional test conditions and coagulant chemicals. Preparation of the water to be tested consisted of spiking Glendale, Calif., water with 50 or 100 µg/L Cr(VI) and adjusting the pH to the desired initial value. One-liter samples were tested in a Phipps and Bird (Richmond, Va.) jar tester. The desired amount of Fe(II) was added to the jar at the beginning of the test and mixed at approximately 40 rpm for 120 min. Previous experiments had shown that 120 min was needed to completely reduce Cr(VI) to Cr(III) at the lowest weight ratio of Fe(II) to Cr(VI) tested. When the jar test was completed, aliquots were removed from the jar and passed through filters ranging in size from 0.45 microns to 1,000 daltons. The filtrate was analyzed for total chromium with an ICP–MS. The polymer used was Magnafloc LT26 (Ciba Specialty Chemicals Corporation). (Magnafloc 1820A was recommended to the authors, but a company representative reported that this polymer was unavailable and recommended the use of Magnafloc LT26 in its place.)

Table 3.12
Summary of the experimental plan for reduction and precipitation studies

Phase	Objective	Water	Reductant	Variables
A	Evaluate iron reduction kinetics	Synthetic	Fe(0) Fe(II)	Time, pH, reductant dosage
B	Impact of co-occurring parameters on reduction kinetics	Synthetic	Fe(0) Fe(II)	Time, reductant dosage, co-occurring parameter
C	Simulated total chromium removal	Synthetic	Fe(II)	Time, reductant dosage, filter pore size
D	Evaluate reduction kinetics and simulated total chromium removal in natural water supplies	LADWP, Glendale	Fe(II)	Time, reductant dosage, pH, filter pore size

Cr(VI) Reduction Experiments in Synthetic Water Supplies

Kinetics of Cr(VI) Reduction by Fe(II) in the Presence of Various Ions

The experiments using Fe(II) as the reductant were operated under the following conditions:

- Initial Cr(VI) concentration: 100 µg/L
- Ion concentrations: 100, 250, and 400 µg/L of phosphate, sulfate, silica, and arsenate
- Ionic strength: 0.1 M adjusted with potassium chloride (KCl)
- Fe(II) dosage: $FeCl_2 4H_2O$ (Fisher), 1,000 µg/L as Fe(II), weight ratio of (Fe(II):Cr(VI) = 10:1
- pH: 5.5, 6.5, and 7.5
- Alkalinity: 50 mg/L as $CaCO_3$

Chemical equilibrium computer software (MINEQL) was used to study the fate of Fe(II) after it had been used to reduce Cr(VI) to Cr(III). The MINEQL program was developed at the Massachusetts Institute of Technology in the mid-1970s to calculate chemical equilibria for water supplies by using numerical algorithms.

Kinetics of Cr(VI) Reduction by Fe(0)

The experiments using Fe(0) as the reductant were operated under the following conditions:

- Initial Cr(VI) concentration: 100 µg/L
- Ionic strength: 0.1 M adjusted with KCl
- Fe(0) dosage: zerovalent iron (99.9% powder, Aldrich), 1,000 µg/L, weight ratio of Fe(0):Cr(VI) = 10:1
- pH: 4.5, 5.5, 6.5, and 7.5
- Alkalinity: 50 mg/L as $CaCO_3$

Cr(VI) Reduction and Total Chromium Removal Experiments in Synthetic and Natural Waters

These experiments compared the performance of the reduction and precipitation process at reducing Cr(VI) and removing total chromium from synthetic and natural waters. Filters and polymers used in the experiments were:

- Filters: 0.45 µm (nylon, Fisher), 0.2 µm (nylon, Fisher), 0.02 µm (nylon, Whatman), 100 KD (YM 100, Amicon Inc.), 10 KD (YM 10, Amicon Inc.), 1 KD (YM1, Amicon Inc.)
- Polymer: 0.5 and 1.0 mg/L Magnafloc LT26 (Ciba Specialty Chemicals Corporation)

CHAPTER 4
CHROMIUM OCCURRENCE AND CO-OCCURRENCE ANALYSIS

The project's literature review determined that limited data existed on the occurrence of chromium in drinking water sources. Little information was available on such basic topics as at what concentration and in what speciation chromium occurs in drinking water sources. Also, few data existed regarding the composition of the water quality matrix in which chromium occurs or whether other contaminants that might require treatment are commonly present in water supplies containing chromium. The authors concluded that a limited study of chromium occurrence would benefit the project and the drinking water community in general.

OBJECTIVES OF THE CO-OCCURRENCE ANALYSIS

The objective of this analysis was to perform a first-order estimate of chromium occurrence and to determine at what level other constituents of concern occur in the presence of chromium. In a narrow sense, the authors thought this analysis would be useful in developing test water supplies with which to conduct bench-scale evaluations. In a broader sense, they believed that an understanding of chromium occurrence and co-occurrence would help them to determine key parameters that might influence potential treatment processes, to evaluate potential opportunities for process synergy through co-removal of other contaminants, and to determine whether multiple treatment processes may be required to treat chromium-contaminated water.

SOURCE DATA

A basic requirement was for the analysis to be performed retrospectively—that is, by using existing data. Several existing databases on occurrence were investigated with respect to their suitability for the project, and it was determined that the National Water Information System (NWIS) database created and maintained by the U.S. Geological Survey (USGS) provided the most comprehensive source of water quality data for the analysis. NWIS is a compilation of water quality data from approximately 1.5 million sites across the United States. Although the NWIS database is accessible online, the available online tools were not suitable for the complex queries and large-scale aggregation of data envisioned for this task. For this reason, a copy of the NWIS data was obtained through a third-party reseller, EarthInfo Inc. EarthInfo supplied the data in a set of eight CD-ROMs, which included the complete NWIS data set for surface water and groundwater supplies up to 1998. Once the data were obtained from EarthInfo, they were transferred into a database developed by MEC and residing on a MySQL 4.0.1 database server. In this form, the data could be accessed as an ODBC (open database connectivity) data source using both Statistica® and Microsoft Access®.

Although the NWIS data set was deemed the most suitable for the analysis, some basic limitations are inherent in using it. First, the data residing in NWIS represent hundreds, if not thousands, of surveys conducted over the past 50+ years. Though it is unknown how many, if any, of these surveys specifically focused on chromium, it is safe to assume that in practically all cases chromium was not the focus of the sampling program; it was more likely one of a number of parameters measured during sampling. Second, the NWIS data represent a variety of surface water and groundwater supplies, many of which are not suitable as drinking water sources. As discussed later, specific criteria were developed to ensure that the co-occurrence analysis was limited to water supplies whose overall quality made them suitable as sources of potable water. Third, several analytical methods with varying detection levels were used to measure the chromium concentrations recorded in NWIS. Thus, the co-occurrence analysis had to ensure that neither elevated detection levels nor out-of-date analytical techniques unduly influenced results. Fourth, although the NWIS data provide geographically and geologically diverse sampling locations, the locations are not uniformly distributed throughout the United States. Typically, sampling locations are grouped within political boundaries (states), and some states are over-represented whereas others are under-represented. This may simply reflect the fact that one state has collected a large amount of groundwater quality data and another state has not. Ultimately, this analysis has the drawback of any retrospective analysis—it was limited to analyzing data that were generally collected for different purposes.

CREATING THE CHROMIUM GROUNDWATER CO-OCCURRENCE DATABASE

Because NWIS contains a large amount of water quality data, a process had to be developed for extracting specific data that would be useful in this project. A four-step process was developed to extract data from NWIS and to create the Chromium Groundwater Co-Occurrence Database. This process is illustrated in Figure 4.1. The first step involved querying the NWIS database and extracting all the groundwater stations for which chromium data existed along with the chromium values recorded for them. Once the stations were identified, the values for co-occurring parameters of interest were also extracted for each station. Second, the mean values of these parameters were tested against the criteria listed in Table 4.1. If the values did not fit within the pre-established criteria, these water supplies were assumed to be unsuitable as drinking water sources and these sites were not included in the chromium database. The wide ranges of data shown in Table 4.1 were set to allow incorporation of as much chromium data as deemed reasonable. Third, detection levels were considered for each parameter. The detection level for each parameter was tested against the criteria in Table 4.1. If the detection level exceeded the "acceptable" level, the data point was dropped. Last, the remaining data were placed in a Microsoft Access database from which values could be extracted for statistical analysis.

If the value for a sample was recorded as zero (0) and no detection level was indicated, that value was not used for statistical analysis of the data. If the sample was listed as a nondetect or 0 and a detection level within the acceptable limit was available, the value of the nondetect was estimated as 0.5 times the detection level for purposes of the statistical analysis.

Using the methods just described, two databases were developed—the first devoted to total chromium and the second to Cr(VI). The Total Chromium Groundwater Occurrence Database contains approximately 180,000 entries relating total chromium occurrence and co-occurrence data for 9,364 groundwater sites located in 49 of the 50 states. The smaller Hexavalent Chromium Groundwater Occurrence Database contains approximately 35,000 entries relating Cr(VI) occurrence and co-occurrence data for 1,282 groundwater sites located in 24 of the 50 states.

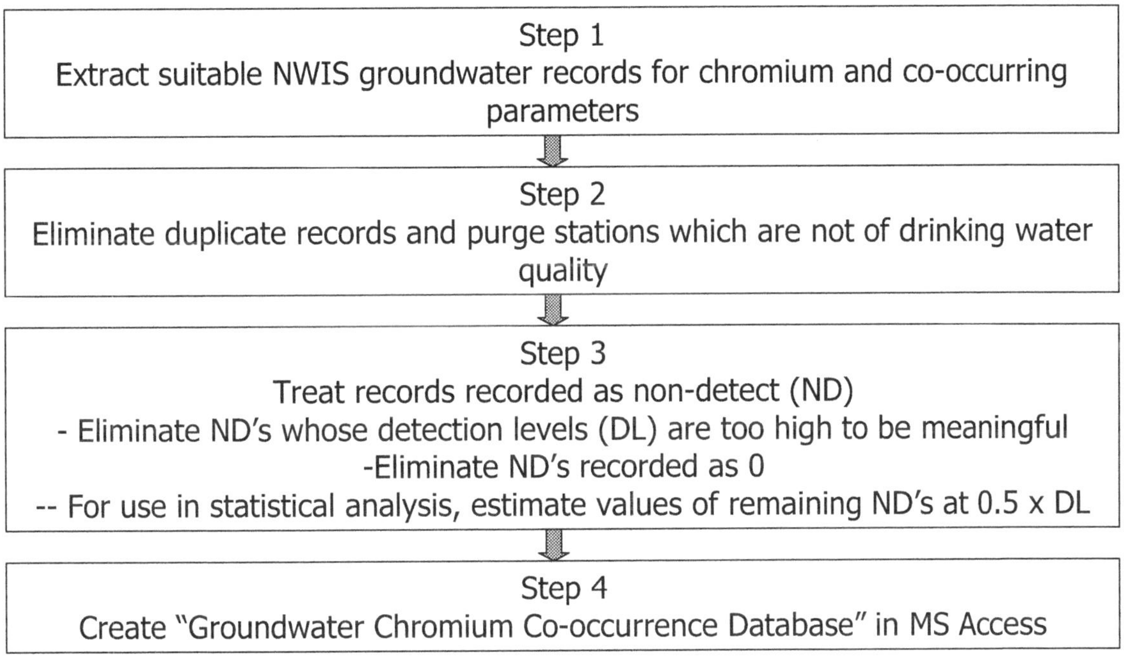

Figure 4.1 Database development flow chart

CHARACTERISTICS OF CHROMIUM OCCURRENCE

Based on the NWIS data extracted by the methods described, the mean total chromium concentration of 9,364 potable-quality groundwater supplies throughout the United States was 8.2 μg/L. The mean Cr(VI) concentration of 1,654 potable-quality groundwater supplies throughout the United States was 4.9 μg/L. These values are far lower than the current USEPA MCL of 0.1 mg/L for total chromium.

The impact of a revised chromium MCL on the drinking water profession can be estimated by developing a cumulative probability distribution of chromium occurrence. Assuming that the probability of chromium occurrence is a continuous function, it can be represented by the cumulative probability function

$$F(X) = \int_{L}^{U} f(x)d(x) \qquad 4.1$$

in which L and U represent the smallest and highest measured chromium values.

The cumulative probability distribution for total and hexavalent chromium is plotted in Figure 4.2. This distribution curve indicates that less than 1% of the groundwater supplies in the database would require treatment to meet USEPA's current MCL for chromium. A decrease in the MCL to 50 μg/L would have little impact on this number. However, if the total chromium MCL was substantially lowered or if a Cr(VI)-specific MCL was established at a level substantially lower than the current total chromium MCL, more groundwater supplies would require treatment. For example, an MCL of 10 μg/L for total or hexavalent chromium would affect approximately 30% and 15% of the groundwater supplies in the database, respectively.

Table 4.1
Database parameters and data acceptance criteria

Parameter	Unit	Acceptable data range Minimum	Maximum	Acceptable detection limit
Alkalinity	mg/L	2	1000	-
Arsenic	μg/L	0.5	150	10
Calcium	mg/L	0.5	500	1
Chloride	mg/L	0.5	500	10
Fluoride	mg/L	0.1	10	0.1
Iron (dissolved)	μg/L	50	5000	10
Iron (total)	μg/L	50	5000	10
Potassium	mg/L	0.5	50	1
Magnesium	mg/L	0.1	100	0.1
Manganese (dissolved)	μg/L	1	500	10
Manganese (total)	μg/L	2	500	10
Sodium	mg/L	0.3	500	10
Nitrate	mg/L	0.1	50	1
pH	-	0	14	-
Phosphate (total)	μg/L	0.1	500	0.1
Selenium	μg/L	.5	200	10
Silica	mg/L	0.1	150	0.1
Sulfate	mg/L	2	1000	10
TDS	mg/L	10	2000	-
TOC	mg/L	0.5	50	0.1
Temperature	°C	0	50	-
Total hardness	mg/L	10	1000	-
Well depth	feet	N/A	N/A	-
Chromium (total)	μg/L	1	500	10
Chromium (hexavalent)	μg/L	0.2	500	10

For the purpose of this analysis, total chromium and Cr(VI) occurrence was divided into two categories. Sites with mean concentrations of 10 μg/L total chromium or greater were considered high total chromium sites; those with mean concentrations of less than 10 μg/L were considered low total chromium sites. A similar approach was used for Cr(VI), except that the two categories were divided at a mean concentration of 5 μg/L. Sites with mean concentrations of 5 μg/L Cr(VI) or greater were considered high Cr(VI) sites; those with mean concentrations of less than 5 μg/L were considered low Cr(VI) sites. Under these criteria, 2,570 of 9,364 sites (27%) had "high" levels of total chromium, and 372 of 1,282 sites (29%) had "high" levels of Cr(VI). Figures 4.3 and 4.4 plot total chromium and Cr(VI) occurrence by geographic location in the contiguous United States. As a point of comparison and using the same criteria as in this analysis, the California DHS survey (Table 2.1) reported that 10% of the drinking water sources tested contained "high" levels of Cr(VI).

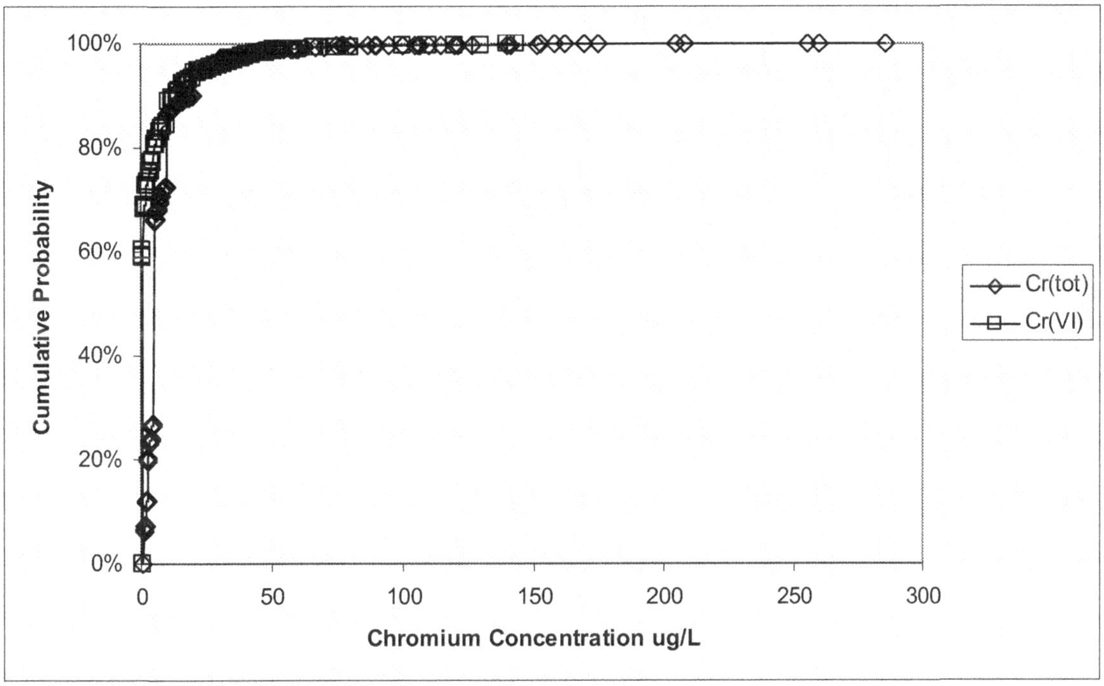

Figure 4.2 Cumulative probability distribution for total and hexavalent chromium in U.S. groundwater supplies

Figure 4.3 shows that the total chromium database contains chromium occurrence data for practically all political divisions and geographic regions of the United States. "High" total chromium sites are present in nearly every state and geographic region. This indicates that on a national basis, total chromium is present in groundwater at concentrations exceeding 10 μg/L. Thus the occurrence of total chromium is national, not regional, in scope.

Conclusions regarding the occurrence of Cr(VI) are less clear. The Cr(VI) database contains fewer stations distributed among fewer states. The absence of chromium data in a particular region or state does not mean chromium is not present there; it merely indicates that no data meeting the criteria listed in Table 4.1 were available in NWIS. Nonetheless, sites with "high" Cr(VI) concentrations are located in the Southeast (Florida), the Northeast (Connecticut, New York, and Pennsylvania), the north central (Michigan) and south central (Louisiana) regions, the Southwest (Arizona), and the Northwest (Washington). Like total chromium, Cr(VI) appears to occur in groundwater supplies throughout the nation.

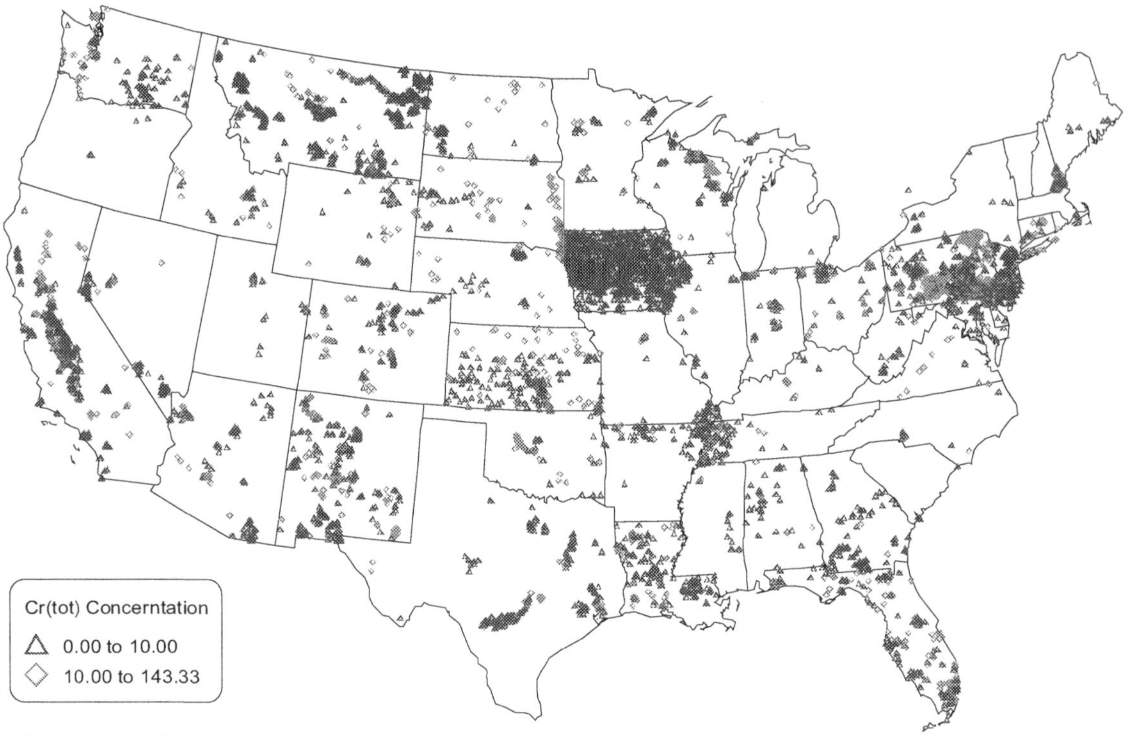

Figure 4.3 Total chromium occurrence in the contiguous United States

Figure 4.4 Hexavalent chromium occurrence in the contiguous United States

CHARACTERISTICS OF CHROMIUM CO-OCCURRENCE

The water quality matrix in which chromium occurs is of prime importance to the evaluation, selection, and ultimate performance of any chromium treatment method. Experience gained by the authors in treating the inorganic contaminant arsenic has indicated that treatment methods for inorganic contaminants can be quite sensitive to the presence of other contaminants. For example, the presence of sulfate, phosphate, silica, or TDS, or the pH at which treatment occurs, can influence the effectiveness of arsenic treatment. Similarly, these or other water quality parameters could influence chromium treatment. In addition, other inorganic contaminants of concern such as arsenic, nitrate, selenium, uranium, vanadium, or fluoride may co-occur with chromium. Therefore, the need for co-removal may be important when a utility selects a treatment technology.

The co-occurrence analysis asked three questions, summarized as follows:

- What are the general water quality characteristics of groundwater supplies containing low and elevated concentrations of total or hexavalent chromium?
- Are the water quality characteristics of groundwater supplies that contain elevated concentrations of total or hexavalent chromium fundamentally different from those that do not?
- Does the presence of chromium correlate with other water quality parameters?

The general water quality characteristics of groundwater supplies containing low and elevated concentrations of total or hexavalent chromium are summarized in Tables 4.2 and 4.3. The tables are divided into two parts, reflecting the categories in which "high" total and hexavalent chromium concentrations are assumed to be equal to or greater than 10 μg/L and 5 μg/L, respectively. The data in Tables 4.2 and 4.3 allow a "typical" water supply containing elevated concentrations of total or hexavalent chromium to be described. For example, median co-occurrence concentrations can be used to characterize a "typical" supply. A "typical" water supply containing elevated concentrations of total chromium has an alkalinity of 200 mg/L as $CaCO_3$, a pH of 7.4, a TDS concentration of 300 mg/L, a sulfate concentration of 30 mg/L, and a silica concentration of 16.8 mg/L as SiO_2. In contrast, a water supply representing a 75th-percentile co-occurrence has an alkalinity of 336 mg/L as $CaCO_3$, a pH of 7.7, a TDS concentration of 576 mg/L, a sulfate concentration of 119 mg/L, and a silica concentration of 26 mg/L as SiO_2.

Tables 4.2 and 4.3 illustrate that the quality of groundwater supplies containing elevated concentrations of total or hexavalent chromium is not appreciably different from that of supplies containing little or no chromium. In order to investigate this conclusion further, the cumulative probability distribution for each co-occurring parameter was developed and plotted. These plots compared the distributions of each parameter in the "low" and "high" chromium supplies. No significant difference in distribution was observed for any parameter. As an example, Figure 4.5 compares the distributions of pH in supplies with "low" and "high" total chromium concentrations. Appendix A contains additional plots. As shown in this figure, there appears to be no fundamental difference in the pH of water supplies that contain elevated concentrations of total chromium and those that do not.

Table 4.2
Characteristics of groundwater supplies containing total chromium (in mg/L, alkalinity in mg/L as $CaCO_3$, Si as mg/L SiO_2)

	Low Total Chromium (Cr(tot) < 10 ug/L)							High Total Chromium (Cr(tot) >= 10 ug/L)							
Parameter	Valid N	Mean	Median	Minimum	Maximum	Lower Quartile	Upper Quartile	Valid N	Mean	Median	Minimum	Maximum	Lower Quartile	Upper Quartile	Parameter
CRTOT	6794	4.2	5.0	1.0	9.9	2.5	5.0	2570	18.8	10.0	10.0	285.0	10.0	20.0	CRTOT
WELLDEPT	5855	306.1	135.0	0.0	6006.0	55.0	305.0	1984	257.9	125.0	6.0	6271.0	53.0	275.0	WELLDEPT
ALK	3727	238.8	213.0	2.1	2500.0	115.0	302.0	1582	273.8	200.0	3.0	1087.9	105.0	336.0	ALK
ARS	351	4.0	2.0	0.5	90.0	1.0	3.5	254	5.4	2.0	0.5	130.0	1.0	4.0	ARS
CA	6551	77.5	61.5	0.5	500.0	25.0	100.0	2412	97.9	53.0	0.5	500.0	21.0	114.1	CA
CL	6341	38.5	13.3	0.5	500.0	5.0	36.0	2292	49.7	16.2	0.5	500.0	5.5	48.9	CL
F	6177	0.5	0.3	0.1	10.0	0.1	0.5	2337	0.5	0.2	0.1	9.8	0.1	0.6	F
FEDISS	3757	881.8	400.0	50.0	5000.0	120.0	1264.0	1497	766.0	270.0	50.0	5000.0	90.0	1100.0	FEDISS
FETOT	1215	1076.6	600.0	50.0	5000.0	206.7	1600.0	367	1126.4	570.0	50.0	5000.0	160.0	1800.0	FETOT
K	5635	4.9	3.1	0.5	50.0	1.9	5.8	1955	6.2	3.7	0.5	50.0	2.0	7.7	K
MG	6340	21.6	16.0	0.1	100.0	6.3	31.2	2154	20.9	12.0	0.1	100.0	4.8	28.0	MG
MNDISS	4937	70.6	27.5	1.0	500.0	7.0	92.0	1830	73.9	30.0	1.0	500.0	10.0	90.0	MNDISS
MNTOT	825	99.8	40.0	2.0	500.0	10.0	158.0	296	107.7	50.0	5.0	500.0	20.0	160.0	MNTOT
NA	6308	61.9	21.0	0.5	500.0	8.2	69.0	2166	73.9	24.2	0.3	500.0	7.5	88.0	NA
NO3	2003	6.1	1.5	0.1	49.0	0.4	8.0	731	7.5	1.8	0.1	49.0	0.3	9.7	NO3
SI	6171	20.4	17.0	0.3	150.0	11.0	25.3	2044	20.7	16.8	0.1	112.5	11.0	26.0	SI
PO4	56	0.7	0.2	0.1	9.5	0.1	0.5	31	0.5	0.2	0.1	2.8	0.2	0.6	PO4
TOC	920	6.6	3.4	0.5	49.0	1.3	8.6	657	6.8	3.3	0.5	50.0	1.6	8.9	TOC
SE	3992	2.7	1.0	0.5	140.0	0.5	5.0	1108	3.9	0.6	0.5	190.0	0.5	2.0	SE
TOTHARD	3437	274.9	247.5	10.0	1000.0	110.0	381.6	1706	245.2	170.0	10.0	1000.0	78.0	337.3	TOTHARD
SO4	5989	115.5	38.0	2.0	1000.0	13.8	117.4	2045	127.9	30.0	2.0	1000.0	11.0	119.0	SO4
TDS	2714	413.5	291.0	11.0	2000.0	158.0	535.0	1215	446.4	300.0	18.0	2000.0	166.0	576.0	TDS
TEMP	6004	15.2	13.5	1.2	48.5	11.2	18.8	2143	15.8	14.5	1.5	48.9	11.8	19.0	TEMP
PH	5161	7.3	7.5	2.4	11.4	7.1	7.8	1693	7.2	7.4	2.6	11.0	6.8	7.7	PH

Table 4.3
Characteristics of groundwater supplies containing hexavalent chromium (in mg/L, alkalinity in mg/L as $CaCO_3$, Si as mg/L SiO_2)

	Low Hexavalent Chromium (Cr(VI) < 5 ug/L)							High Hexavalent Chromium (Cr(VI) >= 5 ug/L)							
Parameter	Valid N	Mean	Median	Minimum	Maximum	Lower Quartile	Upper Quartile	Valid N	Mean	Median	Minimum	Maximum	Lower Quartile	Upper Quartile	Parameter
CRVI	1282	0.8	0.5	0.5	4.8	0.5	0.5	372	19.0	10.0	5.0	143.3	7.2	20.0	CRVI1
WELLDEPT	1055	361.8	232.0	0.0	6271.0	94.0	550.0	281	582.5	500.0	11.0	1983.0	142.0	990.0	WELLDEPT
ALK	993	151.8	120.0	2.3	8530.0	68.3	180.0	296	135.5	117.0	3.5	878.0	79.5	168.0	ALK
ARS	86	5.4	2.5	0.5	42.0	0.6	8.6	29	16.7	14.0	0.5	60.0	2.5	20.5	ARS
CA	1229	57.6	33.0	0.5	500.0	12.1	63.5	360	50.6	30.5	1.0	436.0	14.8	59.0	CA
CL	1163	51.8	15.5	1.0	484.0	7.1	46.0	332	77.8	33.2	1.5	450.0	13.0	97.4	CL
F	1231	0.7	0.3	0.1	10.0	0.1	0.6	347	1.3	0.5	0.1	10.0	0.2	1.7	F
FEDISS	498	756.2	240.0	50.0	4900.0	96.0	980.0	66	787.7	229.8	50.0	5000.0	80.0	1400.0	FEDISS
FETOT	229	1003.3	480.0	50.0	4800.0	200.0	1400.0	62	619.5	240.0	50.0	2912.7	110.0	751.4	FETOT
K	1105	4.6	3.3	1.0	45.0	2.3	5.5	311	4.8	3.9	1.0	47.3	2.4	5.7	K
MG	1221	14.3	8.5	0.1	100.0	2.9	16.7	343	16.1	9.4	0.1	100.0	3.2	18.0	MG
MNDISS	898	46.1	13.3	1.0	480.0	5.0	52.0	165	28.9	5.0	1.0	360.0	3.5	30.0	MNDISS
MNTOT	202	99.2	60.0	2.5	490.0	20.0	147.5	45	56.8	20.0	5.0	295.0	10.0	70.0	MNTOT
NA	1195	59.5	21.0	0.9	500.0	8.2	65.0	341	92.7	56.0	1.1	470.0	18.0	130.0	NA
NO3	243	7.6	2.1	0.1	49.0	0.4	11.0	115	9.9	6.6	0.1	44.0	1.0	16.0	NO3
PH	1100	7.2	7.6	3.8	10.2	6.5	8.0	138	8.0	8.0	6.4	9.2	7.8	8.3	PH
PO4	7	0.1	0.1	0.1	0.2	0.1	0.2	9	0.3	0.2	0.0	1.2	0.1	0.2	PO4
SE	348	1.6	0.5	0.5	16.2	0.5	2.0	189	2.8	2.0	0.5	30.0	1.0	3.0	SE
SI	1267	22.3	22.0	0.3	89.7	9.6	31.0	348	25.8	25.0	0.1	97.0	17.0	33.5	SI
SO4	1161	85.4	28.0	2.0	990.0	13.7	76.8	348	109.2	52.0	2.0	830.0	22.0	120.0	SO4
TDS	792	260.2	201.3	10.0	1970.0	110.3	308.9	136	306.9	235.1	27.1	1350.0	135.0	363.5	TDS
TEMP	1175	17.8	16.4	6.8	45.9	13.8	20.7	323	23.9	25.0	4.2	44.5	17.5	30.0	TEMP
TOC	200	5.3	2.3	0.5	50.0	1.2	6.4	26	5.7	4.5	0.8	31.5	2.0	7.3	TOC
TOTHARD	672	162.9	118.4	10.0	1000.0	53.3	202.4	309	188.0	120.0	10.0	1000.0	55.0	223.0	TOTHARD

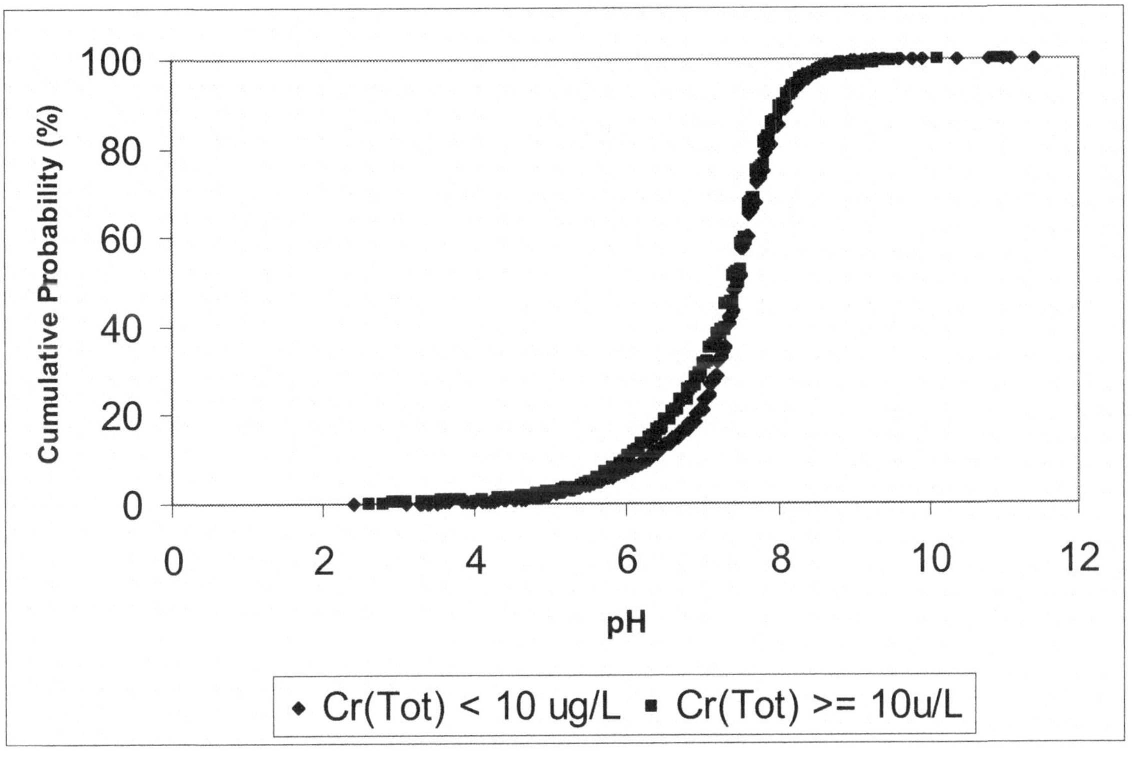

Figure 4.5 Cumulative probability distributions of pH in water supplies containing low and elevated concentrations of total chromium

The strength of the association between two variables and the direction of the association, if any, can be estimated by carrying out a correlation analysis. Assuming that the underlying distribution is normally distributed, the correlation can be estimated by using the Pearson Correlation Coefficient. The Pearson Correlation Coefficient, *r*, is defined as

$$r = \frac{S_{XY}}{S_X S_Y} \qquad 4.2$$

in which S_{XY} is the sample covariance and S_X and S_Y are the standard deviation of the variables. The Pearson Correlation Coefficient can take on values ranging from –1 to +1, with –1 or +1 indicating the greatest relationship and + or – indicating the direction of the relationship.

Six co-occurrence correlation matrixes were created to relate chromium concentration to the concentration of co-occurring parameters. The six matrixes are:

- Total chromium at all sites
- Total chromium at sites with <10 μg/L
- Total chromium at sites with ≥10 μg/L
- Hexavalent chromium at all sites
- Hexavalent chromium at sites with <5 μg/L
- Hexavalent chromium at sites with ≥5 μg/L

Table 4.4 summarizes the *r* values for the correlations. Assuming a 95% confidence level, no statistically significant correlations were found between total or hexavalent chromium and the co-occurring parameters investigated. Thus, the presence of total or hexavalent chromium does not correlate with the presence (or absence) of these parameters.

COMPARISON WITH PROJECT WATER SUPPLIES

As discussed, the quality of "typical" or 50th-percentile water supplies containing chromium can be estimated from the median co-occurrence values. These values, in turn, can be compared with values from the chromium-contaminated natural water supplies studied in this project. Qualitatively, it is possible to assess how these supplies compare with a 50th-percentile chromium-contaminated supply. Table 4.5 compares the median co-occurrence values of natural water supplies containing elevated concentrations of total chromium with those of project water supplies. The concentrations of co-occurring parameters such as TDS and sulfate in the Glendale and LADWP water supplies are greater than those in the 50th-percentile water supplies, and these parameters may adversely affect the performance of treatment technologies such as anion exchange. In general, it appears that the natural water supplies studied in this project will be more difficult to treat than "typical" chromium-contaminated supplies.

Table 4.4
Degree of correlation between chromium and co-occurring parameters, as measured by Pearson Correlation Coefficients at a 95% confidence level

		Sites with Cr(VI)				Sites with Cr(tot)	
Parameter	All sites	< 5 ug/L	>= 5 ug/L	Parameter	All sites	< 10 ug/L	>= 10 ug/L
CRVI1	1	1	1	CRTOT	1	1	1
ALK	-0.04	0.01	-0.14	ALK	0.04	0.00	0.07
ARS	0.02	0.09	-0.26	ARS	0.04	-0.04	0.01
CA	-0.07	0.04	-0.15	CA	0.21	0.08	0.27
CL	0.07	0.01	0.00	CL	0.06	0.04	0.04
CRTOT	0.30	0.10	0.89	CRVI	0.41	-0.22	0.40
F	0.16	0.05	0.08	F	0.03	-0.02	0.08
FEDISS	-0.01	-0.01	-0.06	FEDISS	-0.02	0.07	0.01
FETOT	-0.11	0.00	-0.07	FETOT	-0.02	0.02	-0.07
K	-0.02	0.13	-0.09	K	0.14	0.04	0.17
MG	-0.05	0.10	-0.19	MG	0.07	0.08	0.15
MNDISS	-0.03	-0.03	0.07	MNDISS	0.04	0.07	0.05
MNTOT	-0.10	0.00	0.02	MNTOT	0.08	0.02	0.14
NA	0.10	0.07	0.03	NA	0.06	0.01	0.06
NO3	0.12	0.18	0.13	NO3	0.07	0.01	0.06
PH	0.16	0.15	-0.03	PH	-0.09	0.00	-0.15
PO4	0.19	-0.85	0.03	PO4	-0.10	-0.22	-0.07
SE	0.02	0.11	-0.17	SE	0.05	0.09	0.02
SI	0.04	0.08	-0.08	SI	0.01	0.05	0.01
SO4	0.01	0.02	-0.07	SO4	0.10	-0.01	0.20
TDS	0.00	0.07	-0.12	TDS	0.05	0.06	0.04
TEMP	0.27	0.21	0.08	TEMP	0.05	0.06	0.03
TOC	0.01	0.05	-0.09	TOC	0.11	-0.05	0.19
TOTHARD	-0.06	0.14	-0.20	TOTHARD	0.02	0.08	0.11
WELLDEPT	0.15	0.09	0.05	WELLDEPT	-0.01	0.05	0.03

Table 4.5
Comparison of project water supplies and nationwide U.S. water supplies with total chromium concentrations of ≥10 μg/L

Parameter	Unit	Median for Waters with > 10 ug/L Cr(tot)	Glendale GN3	Glendale GS3	LADWP North Hollywood AT002	LADWP North Hollywood AT004
ALK	mg/L as CaCO3	200	250	180	231	235
ARS	ug/L	2	1.2			
CA	mg/L	53	100	85	91	89
CL	mg/L	16	50	60	27	27
F	mg/L	0.2	0.4	0.3	0.3	0.4
K	mg/L	3.7	3.8	2.3		
MG	mg/L	12	27	29		
NA	mg/L	24	43	39		
NO3	mg/L as N	1.8	7.9	8.8	10.7	9.9
PH	unit	7.4	7.6	7.5	7.6	7.7
SO4	mg/L	30	99	102	64	60.7
SiO2	mg/L	16.8		17		
TDS	mg/L	300	550	400	462	440
TOTHARD	mg/L as CaCO3	170	361	331		

SUMMARY

The occurrence of total and hexavalent chromium in groundwater supplies was estimated by means of a retrospective analysis using water quality data extracted from the NWIS Database. The co-occurrence of other key water quality parameters was estimated as well. Data extracted from NWIS were tested against a conservative set of criteria selected to assure data quality and the suitability of the groundwater for potable consumption. The extracted data were then used to create two databases—the Total Chromium Groundwater Occurrence Database and the Hexavalent Chromium Groundwater Occurrence Database.

Using information from these databases, the mean total chromium concentration for 9,364 groundwater sites whose water quality was suitable for potable consumption was 8.2 μg/L. The mean Cr(VI) concentration for 1,654 groundwater sites whose water quality was suitable for potable consumption was 4.9 μg/L. The occurrence of elevated concentrations of total and hexavalent chromium, defined as 10 μg/L and 5 μg/L, respectively, does not appear to be concentrated in any single geographic region but is distributed throughout the United States.

The water quality characteristics of groundwater supplies containing little total or hexavalent chromium were compared with those of groundwater supplies containing elevated concentrations of total or hexavalent chromium. No meaningful difference in pH, major cations or anions, or contaminants such as arsenic or nitrate was observed between groundwater supplies that contain elevated chromium concentrations and those that do not. It appears that there are no fundamental differences in the composition of groundwater supplies that contain chromium and those that do not.

Correlation matrixes were developed to determine if the presence of total or hexavalent chromium could be directly related to any other contaminant. No correlation was found between total or hexavalent chromium and any of the contaminants investigated.

The median co-occurrence values were used to estimate the composition of a "typical" chromium-containing water supply. A comparison of the composition of this theoretical water supply with the chromium-contaminated groundwater supplies provided to this project by the city of Glendale and LADWP indicated that the utility-provided supplies are likely to be more difficult to treat than a "typical" water supply.

CHAPTER 5
CHROMIUM REDUCTION AND OXIDATION

The chromium redox experiments examined the oxidation of Cr(III) by four oxidants and the reduction of Cr(VI) by three reductants under five water quality conditions (Table 5.1). All experiments were run in triplicate. In the graphs illustrating the results of these experiments, the data point represents the average and the error bars represent one standard deviation of the triplicate experiments.

CONTROL EXPERIMENTS

For quality control purposes, two sets of blank tests were carried out to find out if the diphenylcarbohydrazide used to quantify Cr(VI) would react with any of the oxidants or reductants used in the experiments. First, the appropriate concentration of each reductant and oxidant was dosed with diphenylcarbohydrazide reagent to see if it would cause color formation that would erroneously be quantified as Cr(VI). Second, each oxidant plus Cr(VI) and reagent was compared with only Cr(VI) and reagent to see if the reaction between the oxidant and the reagent would cause any positive or negative interference. Similarly, each reductant plus Cr(III) and reagent was compared with only Cr(III) and reagent to see if the reductant would cause any interference.

Table 5.1
Conditions tested in redox experiments

	Water quality				
	Deionized water	Synthetic water	Modified synthetic water	Reducing water	Glendale water
			Oxidants		
Dissolved O_2	√	√	√	√	√
Cl_2	√	√	√	√	√
Chloramine	√	√			√
$KMnO_4$	√	√	√	√	√
H_2O_2	Not tested owing to analytical interference				
			Reductants		
$SnCl_2$	√	√	√		√
S^{-2}	√	√			√
SO_3^{-2}	√	√			√

Results showed that hydrogen peroxide oxidized the diphenylcarbohydrazide reagent completely so that no red color developed, rendering colorimetric measurement of the Cr(VI) concentration impossible. As a result, hydrogen peroxide was not tested further. Permanganate also reacted with diphenylcarbohydrazide, causing a negative interference. The degree of interference was dependent on the concentrations of permanganate and Cr(VI) (Table 5.2) but was not affected by pH or other water quality characteristics. Thus, a correction factor was developed for each experiment, based on the measured Cr(VI) concentration and the $KMnO_4$ dosage, and applied to all Cr(VI) data. Because the $KMnO_4$ dosage was much greater than the stoichiometric requirement, the $KMnO_4$ concentration was assumed to remain constant for the duration of each experiment, for purposes of determining the correction factor.

The other oxidants (chlorine, chloramines) and all the reductants (stannous chloride, sulfide, and sulfite) showed no interference with the colorimetric method.

For each of the five water quality conditions, a control experiment monitored the stability of Cr(III) or Cr(VI) without any oxidant or reductant. Cr(VI) was stable at all pH levels in all the water quality conditions, including the synthetic water (Figure 5.1), DI water (Figure B.1), Glendale water (Figure B.2), and modified synthetic water (Figure B.3). (Figures B.1–B.17 are located in Appendix B.) In all cases, the total chromium concentration was the same as the Cr(VI) concentration, and there was no difference between filtered and unfiltered samples. Total chromium was not measured in the experiment using DI water (Figure B.1).

Table 5.2
Effect of potassium permanganate on measured Cr(VI)

	Measured Cr(VI) (µg/L)			
$KMnO_4$ dosage (mg/L)	25-µg/L dosage	50-µg/L dosage	75-µg/L dosage	100-µg/L dosage
0	26.1	51.0	76.2	100.5
1	23.5	37.8	64.1	86.1
3	21.5	35.4	55.4	72.8
5	20.8	32.8	53.2	69.7
10	19.4	30.9	37.9	61.7

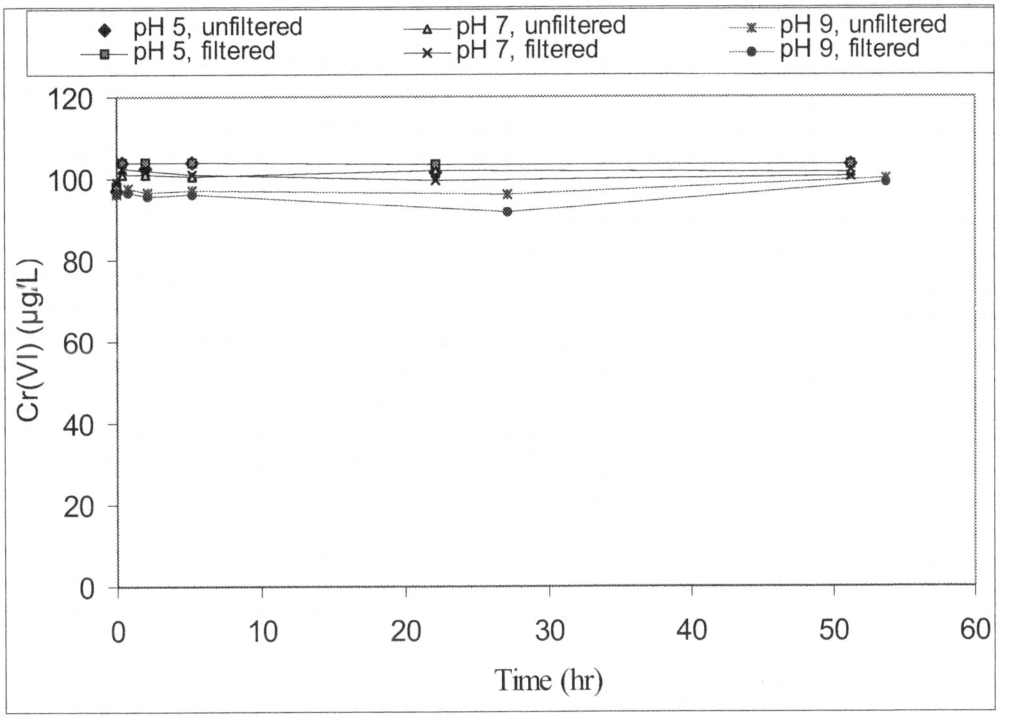

Figure 5.1 Control experiment with 100 µg/L Cr(VI) in synthetic water

Unlike Cr(VI), Cr(III) stability varied with pH and water quality. In the experiments with DI water, no oxidation of Cr(III) to Cr(VI) was observed, based on the measurement of Cr(VI) (Figure B.4). However, total chromium was not measured, so it is not known if any of the chromium was precipitating. In the control experiments with synthetic water, no Cr(VI) was detectable at any pH, so no Cr(III) was oxidized. Cr(III) was stable at pH 5 and 7. However, at pH 9, the unfiltered total chromium remained near 100 µg/L, but total chromium in the filtered sample decreased to nearly 20 µg/L after 50 hours, indicating that 80% of the Cr(III) had precipitated (Figure 5.2). Similar results were observed in the experiments with reducing water and Glendale water (Figures B.5 and B.6). The Glendale water already contained approximately 11 µg/L Cr(VI), so it was supplemented with 90 µg/L Cr(III) for the oxidation experiments or 90 µg/L Cr(VI) for the reduction experiments to bring the total chromium concentration to 100 µg/L. The naturally occurring and added Cr(VI) remained stable (Figure B.6).

One possible explanation for the formation of particulate Cr(III) at high pH is that particulate $CaCO_3$ present in the water might have sorbed some of the Cr(III) and that these particulate species were then removed by the filter. Thus, another control experiment was conducted with the modified synthetic water, which had no particulate $CaCO_3$ (Figure 5.3). However, the same amount of precipitation was observed (comparing filtered and unfiltered samples at pH 9), indicating that the Cr(III) was being lost through precipitation and that the loss was not related to $CaCO_3$.

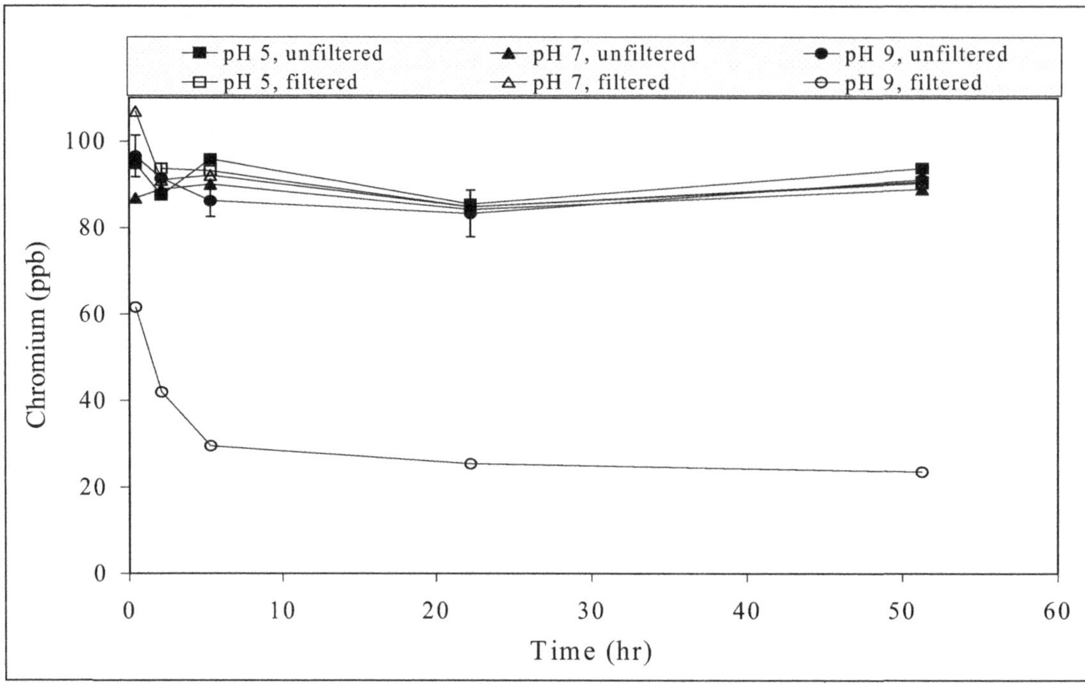

Figure 5.2 Control experiment with 100 µg/L Cr(III) in synthetic water (all measurements were for total chromium)

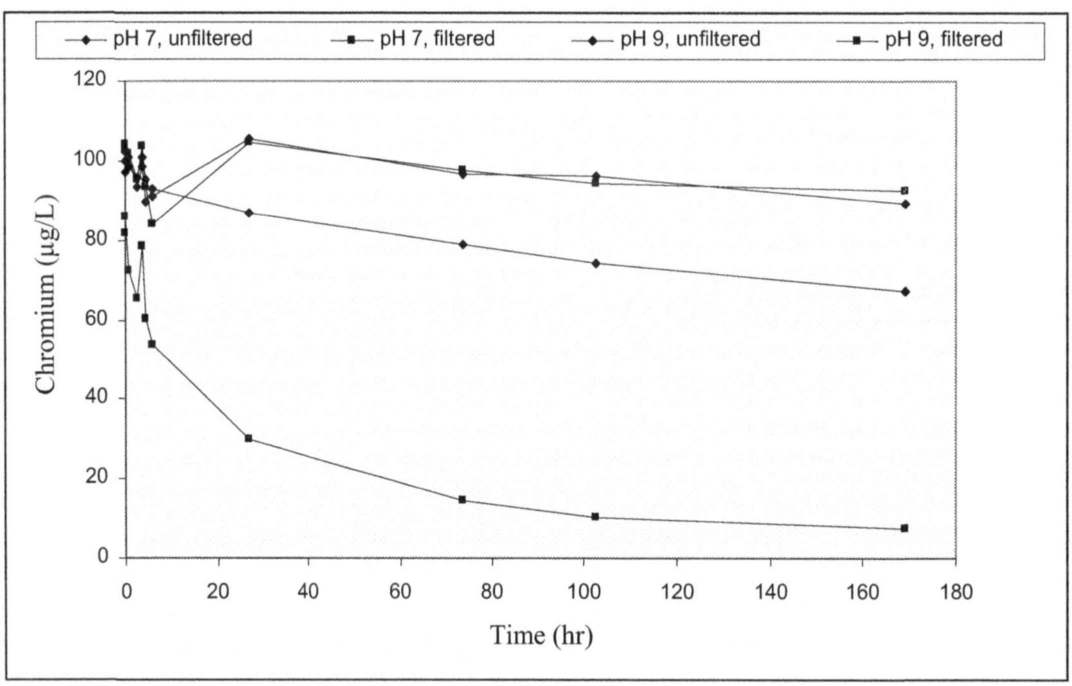

Figure 5.3 Control experiment with 100 µg/L Cr(III) in modified synthetic water (all measurements were for total chromium)

Given the Cr(III) solubility diagram for $Cr(OH)_3$ (Figure 2.2B), it is not surprising that some Cr(III) precipitation might occur at pH 9, although the exact composition and solubility of the solid formed in these tests is unknown. Thus, a set of control experiments were conducted with initial Cr(III) concentrations of 10, 20, and 40 µg/L in synthetic water. A significant amount of Cr(III) precipitation occurred, regardless of the initial Cr(III) concentration, as indicated by the difference between the filtered and unfiltered samples for each concentration (Figure 5.4). Because decreasing the initial Cr(III) concentration would not eliminate formation of the Cr(III) precipitate, and because using a higher chromium concentration was desirable to facilitate chromium measurement, the initial chromium concentrations were maintained at 100 µg/L. However, all subsequent results at pH 9 must be interpreted in light of the possible presence of particulate Cr(III) species.

TRIVALENT CHROMIUM OXIDATION RESULTS

Dissolved Oxygen

All experiments were conducted in water saturated with approximately 7 mg/L of dissolved oxygen (O_2). The Cr(III) control experiments with no oxidant added indicated that dissolved O_2 is not strong enough to oxidize Cr(III) because no Cr(VI) was produced in any of the control experiments (Figures 5.2, 5.3, B.4, B.5, and B.6).

Free Chlorine

The oxidation of Cr(III) by 1 mg/L chlorine (Cl_2) was investigated under all five water quality conditions. The experiments with DI water were only run to 24 hours and showed less than 50% oxidation. A greater degree of oxidation was found at pH 5 and 7 than at pH 9 (Figure 5.5).

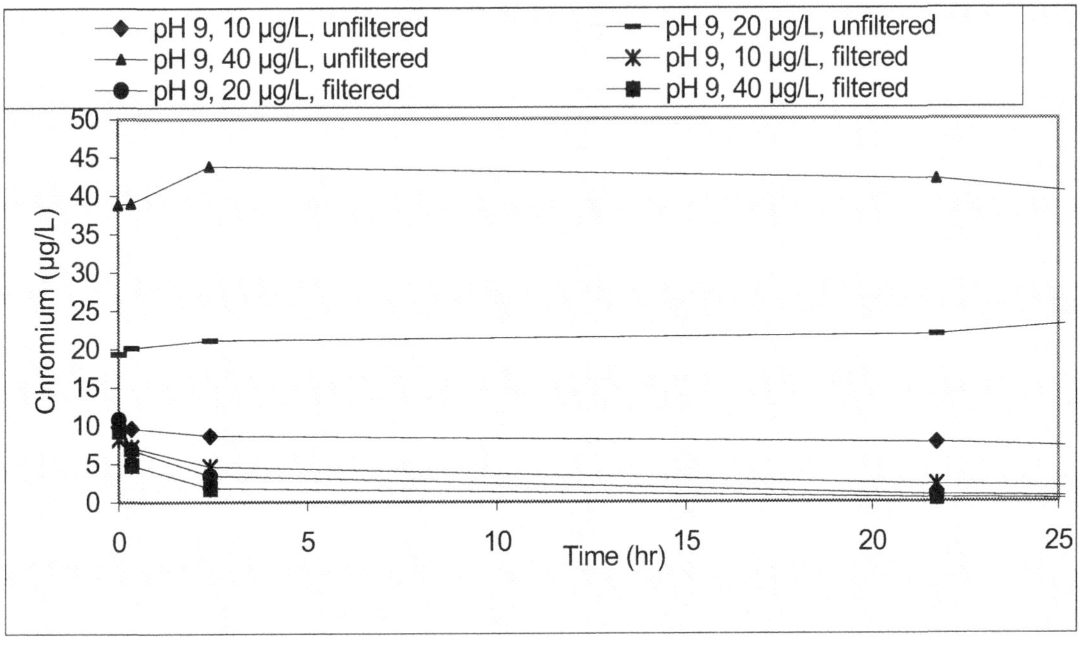

Figure 5.4 Control experiment with 10, 20, and 40 µg/L Cr(III) (all measurements were for total chromium)

The experiments with synthetic water were run to 140 hours (6 days); >90% oxidation was observed at pH 5 and 80% oxidation at pH 7 (Figure 5.6). At both pH 5 and 7, the total chromium concentration remained constant, and the concentrations in filtered and unfiltered samples were the same, indicating no precipitation (Figure B.7 and B.8). At pH 9, very little oxidation was observed. This could be due to the change in chlorine speciation (there is more OCl^- instead of HOCl at higher pH).

Another important factor is the precipitation of the chromium at high pH, which was also observed in the control experiment (Figure 5.2). The chromium concentration in the filtered samples decreased, indicating precipitation (Figure 5.7). The total chromium concentration also dropped significantly, likely because of the precipitated particles not being sampled owing to their size (causing them to be excluded by the pipette tip used to sample) or because the particles settled to the bottom of the bottle. The Cr(VI) concentration approached the total chromium concentration in the filtered samples by the end of the experiment, which likely indicates that the chlorine could not oxidize the precipitated Cr(III). Similar results were observed in the experiments with modified synthetic water (data not shown), indicating that particulate $CaCO_3$ had little impact on the oxidation of Cr(III) by Cl_2.

In addition to these factors, some of the oxidation effects could be explained by reductions in chlorine residual over the experimental time period and variations of chlorine in the reaction bottles. The protocol for the oxidation experiments with free chlorine and chloramines did not include measuring initial chlorine residuals or following chlorine degradation over 24, 96, or 145 hours.

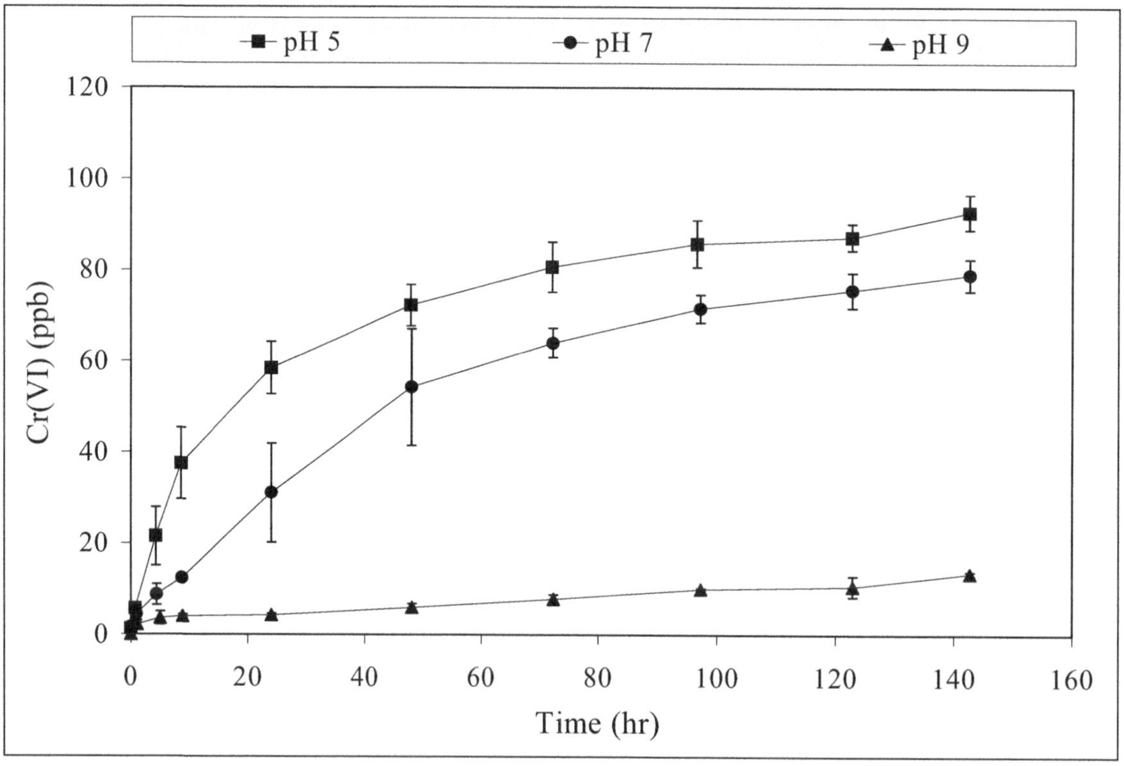

Figure 5.5 Oxidation of 100 μg/L Cr(III) by 1 mg/L Cl_2 in deionized water

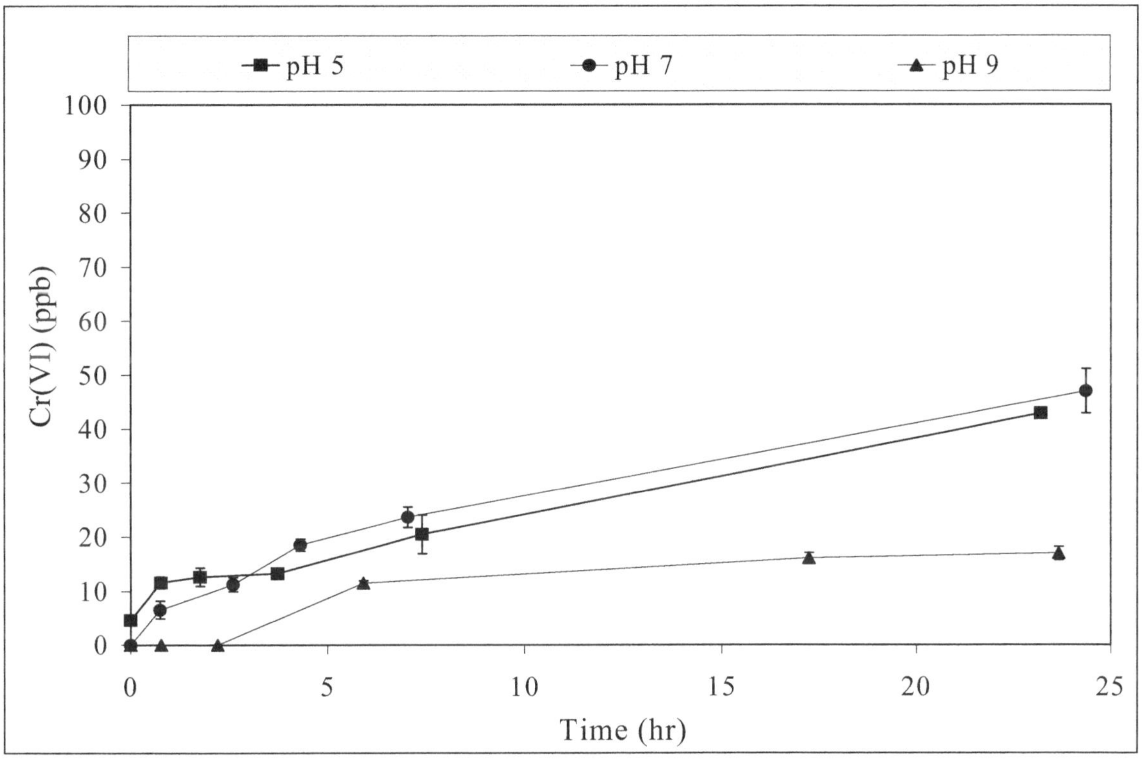

Figure 5.6 Oxidation of 100 μg/L Cr(III) by 1 mg/L Cl_2 in synthetic water

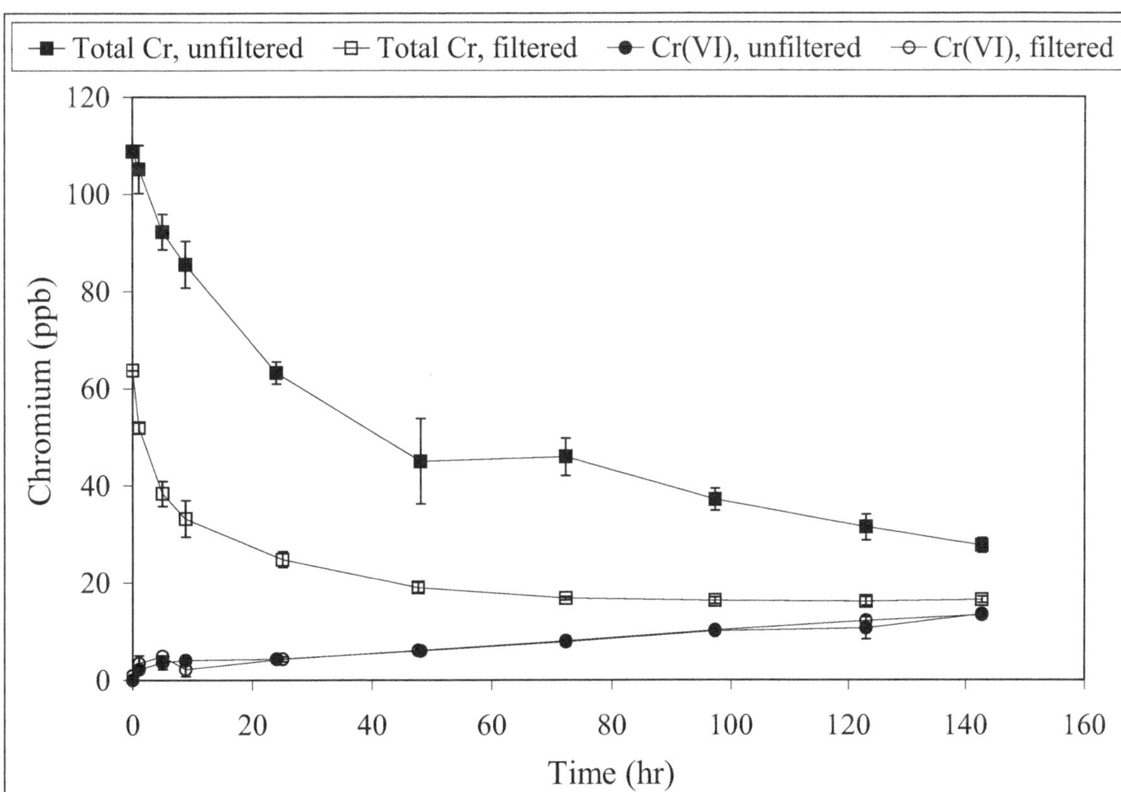

Figure 5.7 Oxidation of 100 μg/L Cr(III) by 1 mg/L Cl_2 in synthetic water at pH 9

In contrast to the DI and synthetic water, oxidation of Cr(III) by Cl_2 in the Glendale water was highest at pH 7 (Figure 5.8). Still, as in the other types of water tested, oxidation at pH 9 was very low. Again, this is likely due to nearly complete precipitation of the Cr(III), which the chlorine was unable to oxidize effectively (Figure 5.9).

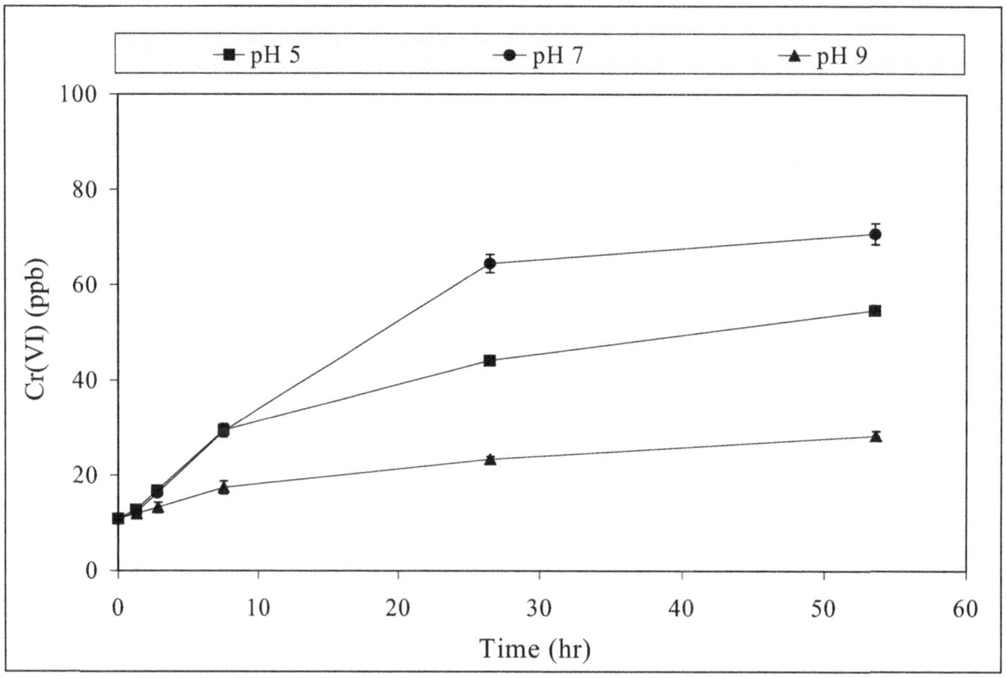

Figure 5.8 Oxidation of 90 μg/L Cr(III) by 1 mg/L Cl_2 in Glendale water

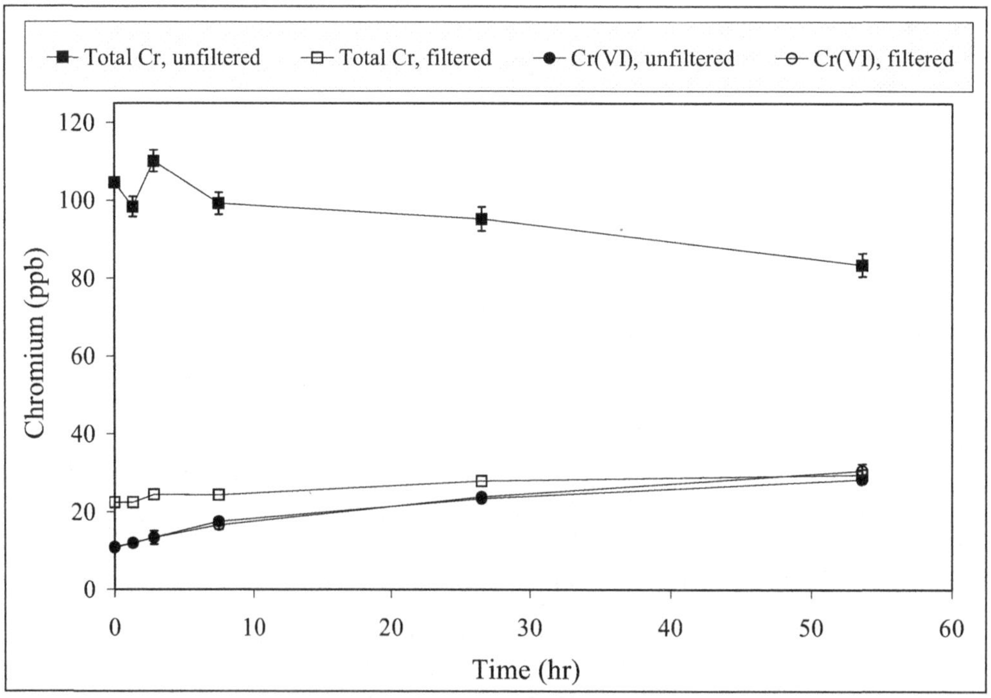

Figure 5.9 Oxidation of 90 μg/L Cr(III) by 1 mg/L Cl_2 in Glendale water at pH 9

Oxidation of Cr(III) by Cl_2 was also investigated at pH 5 and 7 in the reducing water, which was similar to the synthetic water but also contained natural organic matter (NOM) and reduced manganese. Oxidation at pH 9 was not examined, given the small amount of oxidation observed under the other three water quality conditions. In this reducing water, no oxidation of Cr(III) was observed at either pH 5 (Figure 5.10) or pH 7 (Figure 5.11). This may be because the reduced manganese and NOM completely removed the chlorine that had been added, but there are no chlorine residual data from this experiment to confirm this supposition. However, most of the total chromium was precipitated, as indicated by the difference in the filtered and unfiltered samples, so perhaps the Cl_2 was simply unable to oxidize the precipitated Cr(III). It is unclear why such significant precipitation was observed at these low pH values.

Chloramine

In this phase of the testing, oxidation of Cr(III) by chloramine was investigated in the DI and synthetic water. The chloramine dosage was 0.5 mg/L as Cl_2, with a Cl_2:N ratio of 3:1, which should encourage formation of monochloramine. Under the conditions of these experiments, it appears that no oxidation of Cr(III) occurred in the DI water (Figure B.9) or the synthetic water (Figure 5.12). In experiments on synthetic water at pH 5 and 7, there was no difference between the filtered and unfiltered samples (data not shown), but at pH 9, a significant decrease in the chromium concentrations of both unfiltered and filtered samples was observed, indicating precipitation of the Cr(III) (Figure 5.12). This result was similar to the behavior observed in the synthetic water experiments with free chlorine (Figure 5.7). Chloramine also appeared to be ineffective at oxidizing Cr(III) in the Glendale water (Figure 5.13) under the conditions of these experiments. One concern regarding these experiments is that the chloramine residual was quite low and no residual total chlorine concentration was measured initially or after the 96-hour contact time. Preparing low-concentration chloramine doses requires a great deal of attention, and it is possible to create a chloramine dose that is not what was anticipated.

Potassium Permanganate

Oxidation of Cr(III) by $KMnO_4$ was investigated under all five water quality conditions. In DI water, oxidation by 1 mg/L $KMnO_4$ at pH 9 and 7 was nearly complete by the first sampling time (1 hr); oxidation at pH 5 was much slower but was complete by 24 hr (Figure 5.14). This trend was confirmed in experiments with synthetic water, in which oxidation was initially faster at pH 9 and 7; however, with very long detention times, oxidation of Cr(III) to Cr(VI) was ultimately complete at all three pH values (Figure 5.15).

For the synthetic water at pH 5, the effect of $KMnO_4$ dosage was not consistent; 3 mg/L was the most effective (Figure B.10). At pH 7 (Figure B.11) and pH 9 (Figure 5.16), the dosage had less effect, although the 3-mg/L dose was best at some sampling times. For all experiments with $KMnO_4$ in the synthetic water, total chromium remained essentially constant (Figures B.10, B.11, and 5.16), and at pH 5 and 7, total chromium concentrations in the filtered and unfiltered samples were the same. At pH 9, however, significant precipitation of the Cr(III) occurred (Figure 5.17). At pH 9, the filtered total chromium and Cr(VI) concentrations tracked each other and increased with time, perhaps indicating that unlike Cl_2, $KMnO_4$ is able to oxidize particulate Cr(III). Similar results were obtained in the modified synthetic water (data not shown).

The oxidizing ability of $KMnO_4$ was also tested in Glendale water. The 3-mg/L dosage was chosen, given the results with the synthetic water. In contrast to the synthetic water, equilibrium was achieved in Glendale water in several hours, and the amount of oxidation was not as great. However, the same trend of more oxidation at lower pH was observed (Figure 5.18). The small amount of oxidation at pH 9 is likely due to precipitation of the Cr(III) at this high pH, as indicated by the difference between the filtered and unfiltered total chromium samples (Figure 5.19).

A 3-mg/L dose of $KMnO_4$ was also tested in the reducing water at pH 5 and 9 (Figures B.12 and B.13). The degree of oxidation achieved was nearly the same as in the synthetic water and the Glendale water, with more oxidation at pH 5 than at pH 9 and significant precipitation at pH 9. Similar to the Cl_2 tests with reducing water at pH 5 (Figure 5.10), some precipitation of Cr(III) was also observed with $KMnO_4$ in reducing water at pH 5 (Figure B.12). The reason for the Cr(III) precipitation at this low pH is unknown.

Hydrogen Peroxide

This project attempted to evaluate the oxidation of Cr(III) by hydrogen peroxide (H_2O_2), but, unfortunately, peroxide also oxidized the colorimetric reagent used to quantify Cr(VI), so it was not possible to measure the production of Cr(VI). Nevertheless, given that the use of H_2O_2 is listed as a Standard Method (3111B, Section 4a) for oxidation of Cr(III) (APHA, AWWA, WEF 1998), it is likely to be effective.

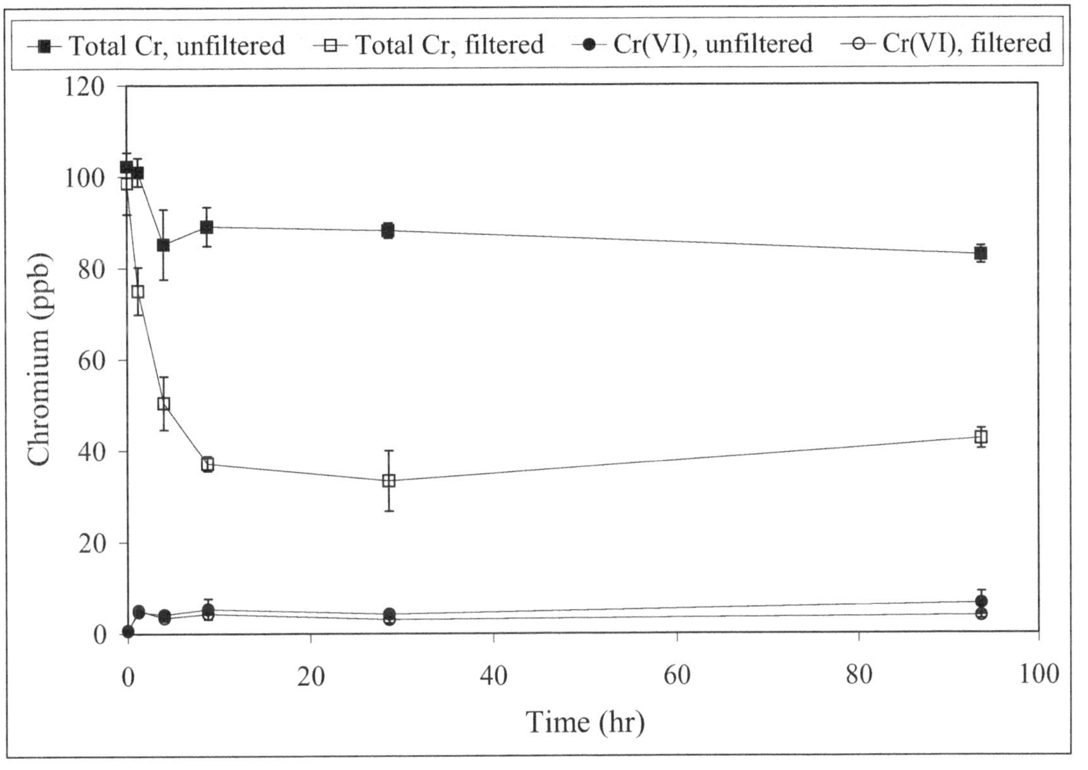

Figure 5.10 Oxidation of 100 µg/L Cr(III) by 1 mg/L Cl_2 in the reducing water at pH 5

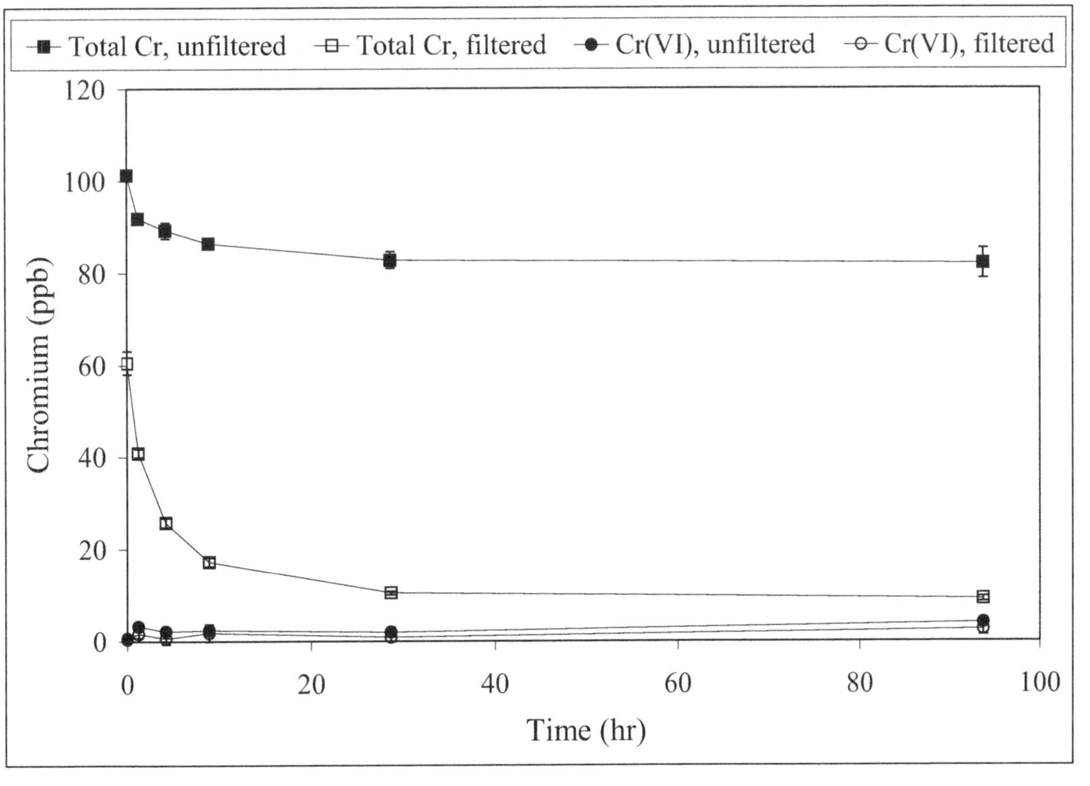

Figure 5.11 Oxidation of 100 µg/L Cr(III) by 1 mg/L Cl_2 in the reducing water at pH 7

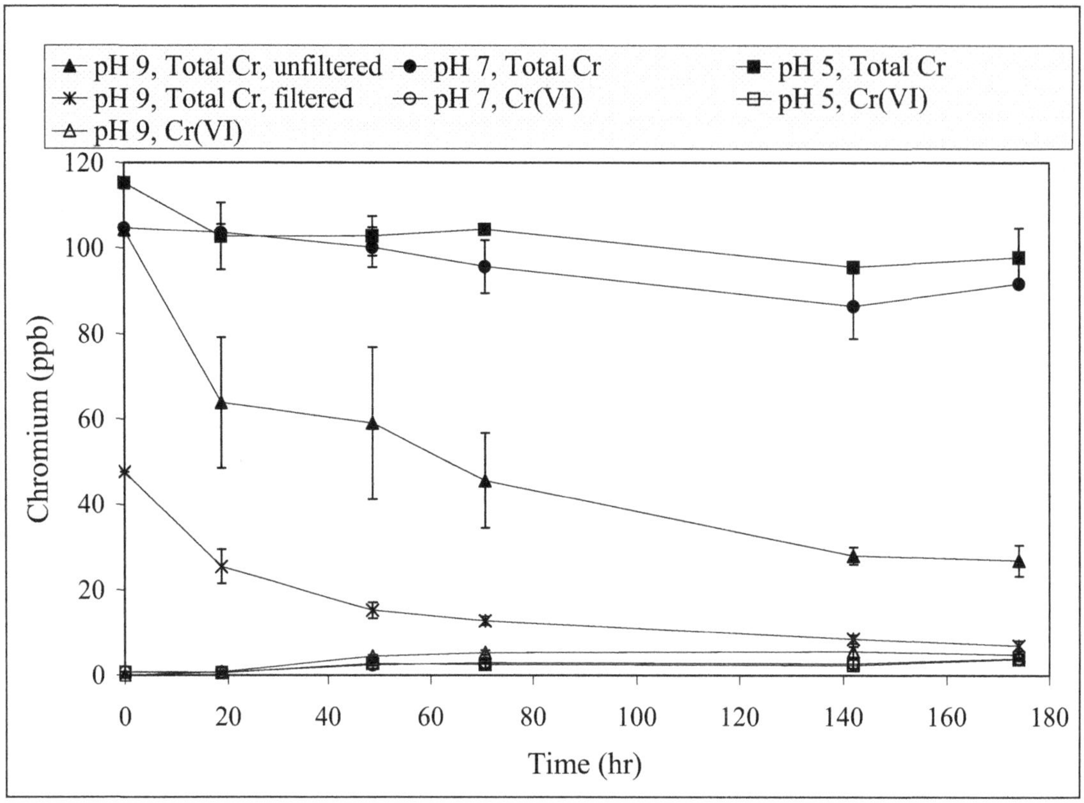

Figure 5.12 Oxidation of 100 μg/L Cr(III) by 0.5 mg/L chloramine in synthetic water

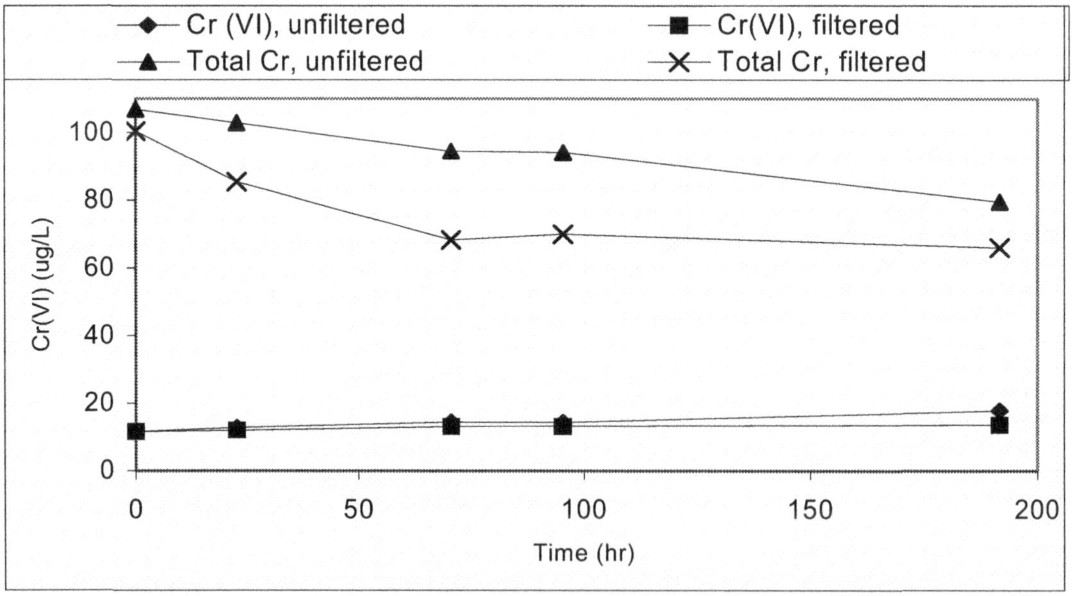

Figure 5.13 Oxidation of 100 μg/L Cr(III) by 0.5 mg/L chloramine in Glendale water at pH 7

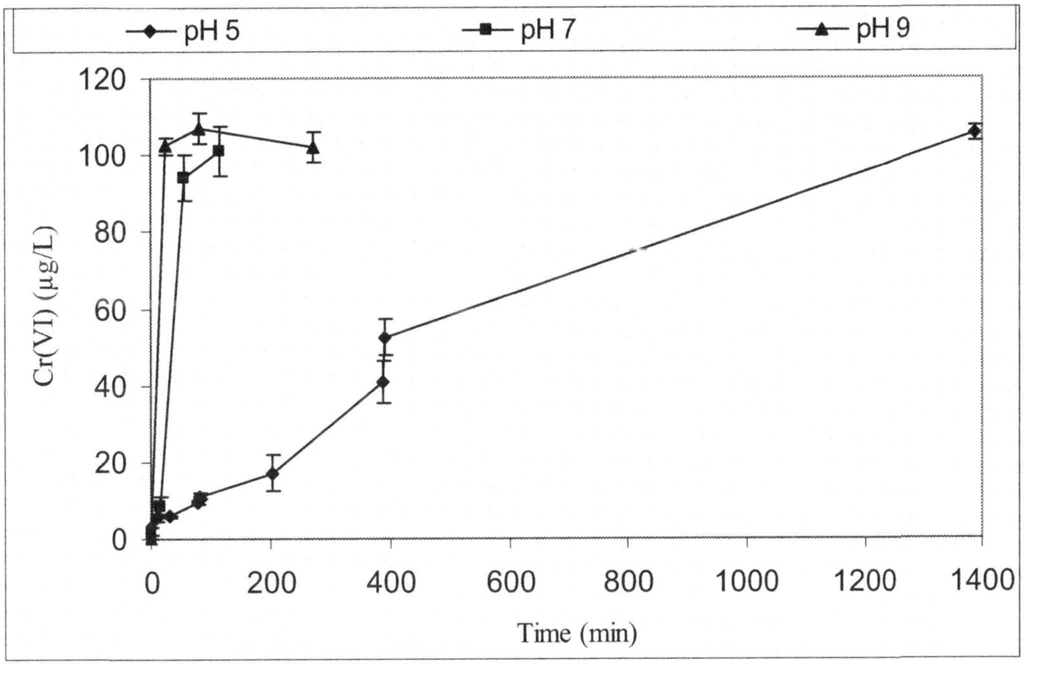

Figure 5.14 Oxidation of 100 μg/L Cr(III) by 1 mg/L $KMnO_4$ in deionized water

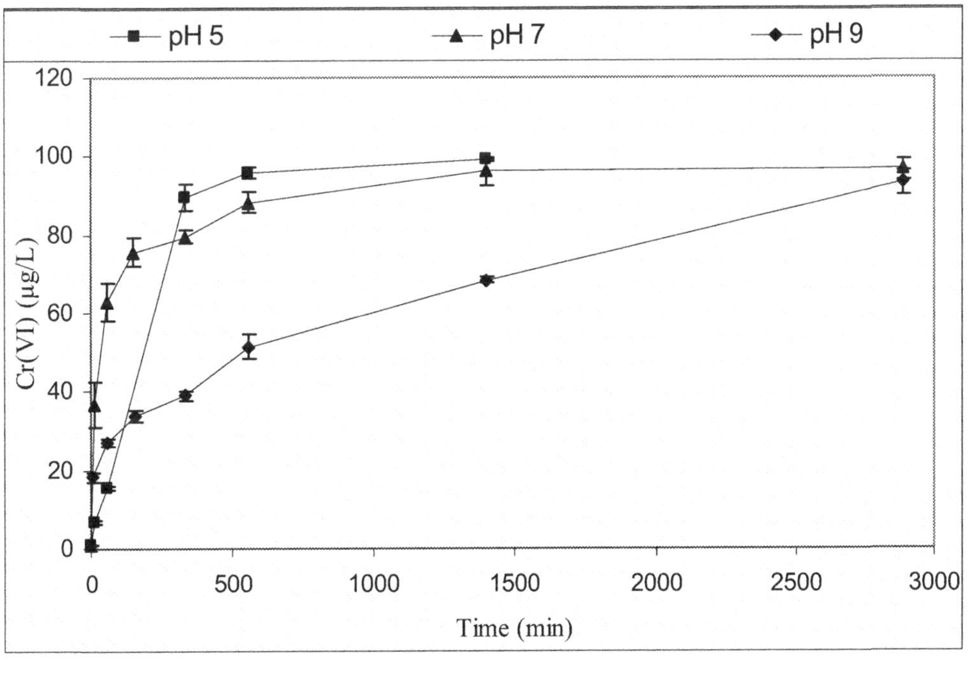

Figure 5.15 Oxidation of 100 μg/L Cr(III) by 1 mg/L $KMnO_4$ in synthetic water

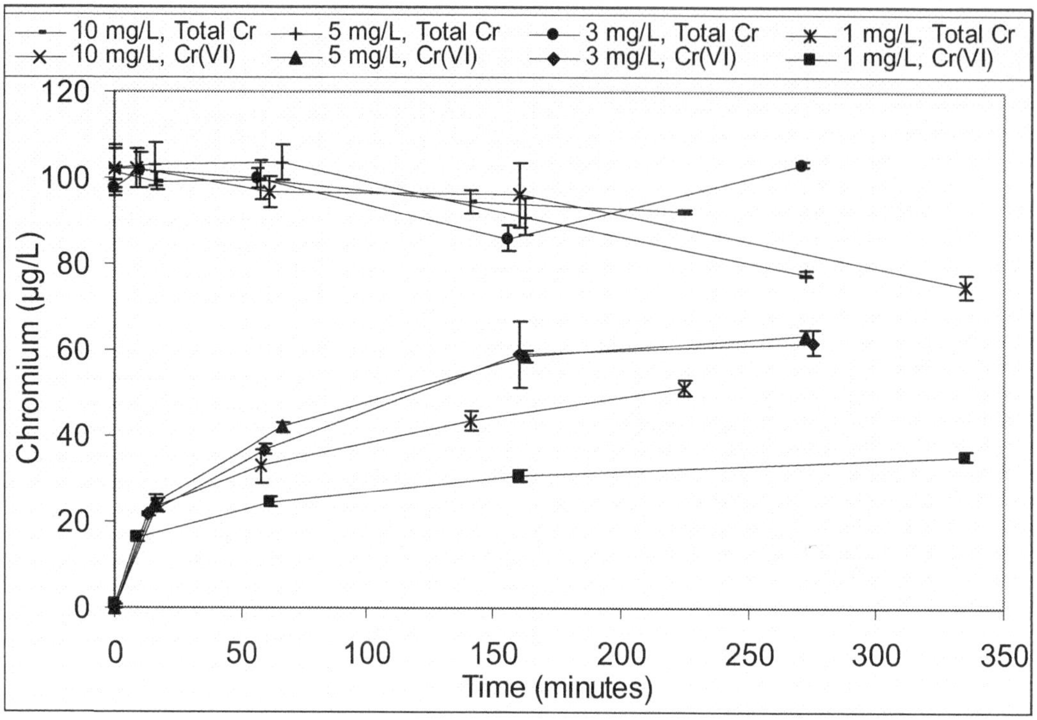

Figure 5.16 Oxidation of 100 µg/L Cr(III) by $KMnO_4$ in synthetic water at pH 9 (samples shown are unfiltered)

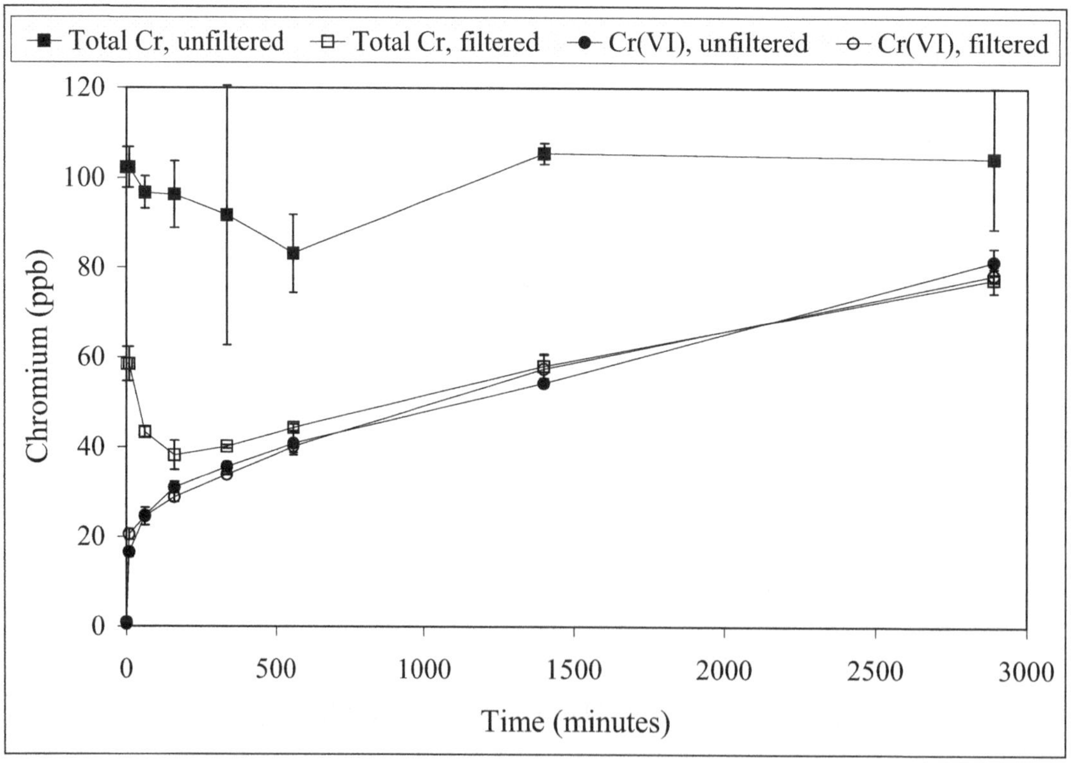

Figure 5.17 Oxidation of 100 µg/L Cr(III) by 1 mg/L $KMnO_4$ in synthetic water at pH 9

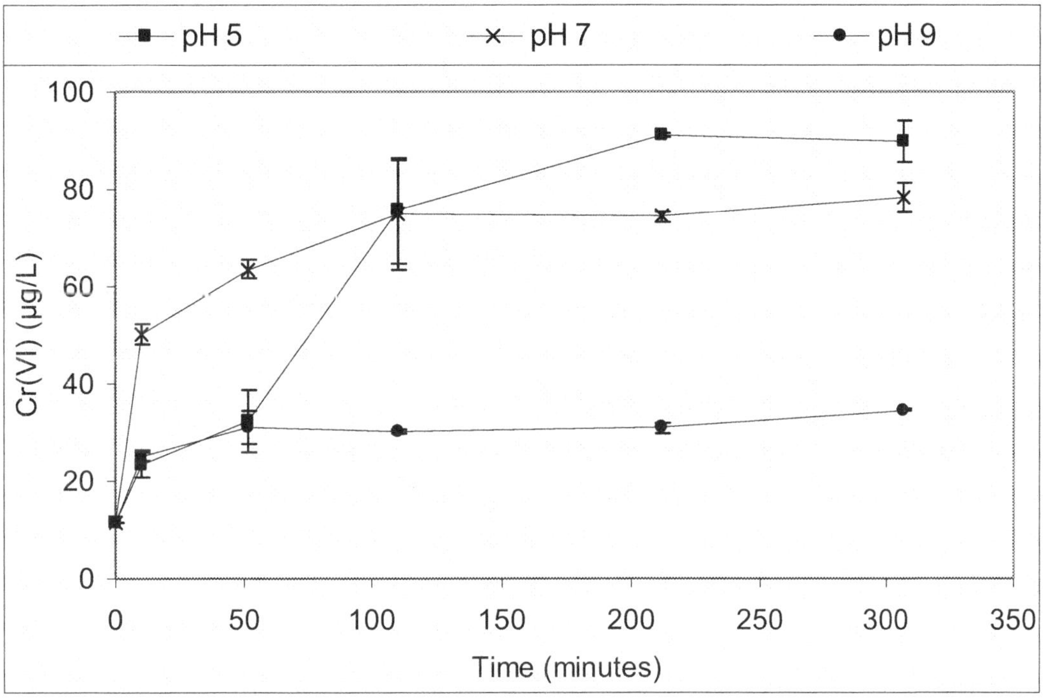

Figure 5.18 Oxidation of 100 µg/L Cr(III) by 3 mg/L $KMnO_4$ in Glendale water

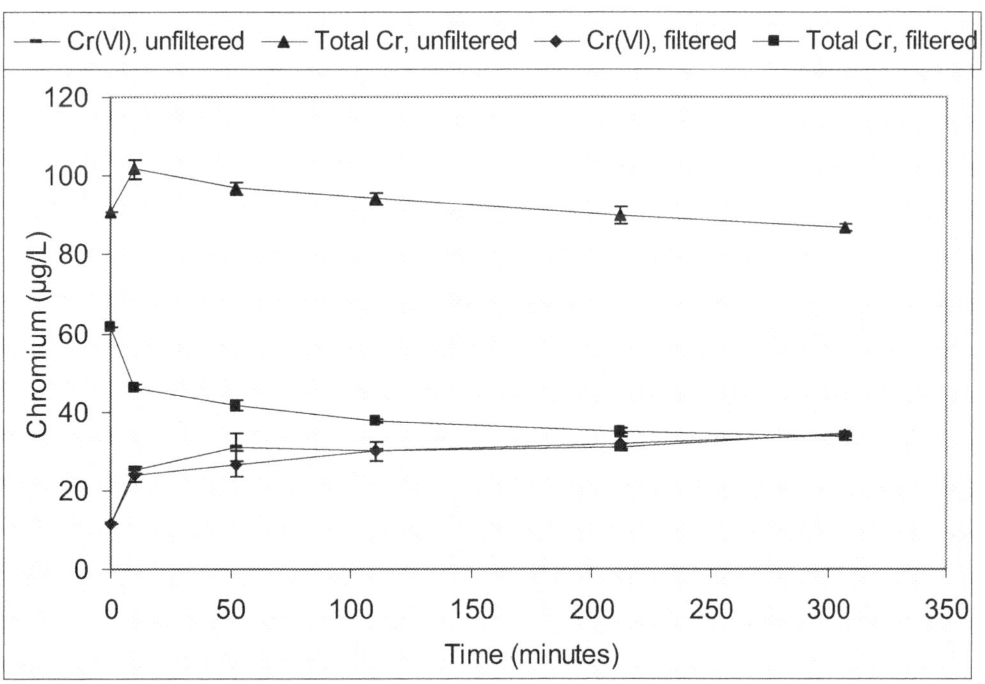

Figure 5.19 Oxidation of 100 µg/L Cr(III) by 3 mg/L $KMnO_4$ in Glendale water at pH 9

HEXAVALENT CHROMIUM REDUCTION RESULTS

Stannous Chloride

The reduction of Cr(VI) by stannous chloride ($SnCl_2$) was investigated in DI, synthetic, and Glendale water, using a dosage of 1.3 mg/L (twice the stoichiometric dosage). In the DI water, reduction of Cr(VI) was complete by the first sampling time (30–40 min). There was no difference between the filtered and unfiltered Cr(VI) samples, but total chromium was not measured (Figure 5.20).

Several interesting phenomena were observed in the experiments with synthetic water. At pH 5, reduction of Cr(VI) was rapid but only 60% complete (Figure 5.21), and there was no difference between filtered and unfiltered Cr(VI) samples. However, total chromium concentrations decreased with time in both unfiltered and filtered samples, and some particulate chromium was formed. Because the Cr(VI) concentration was very close to the filtered total chromium concentration, it appears that as the Cr(VI) was reduced to Cr(III), it precipitated and was removed by the filter.

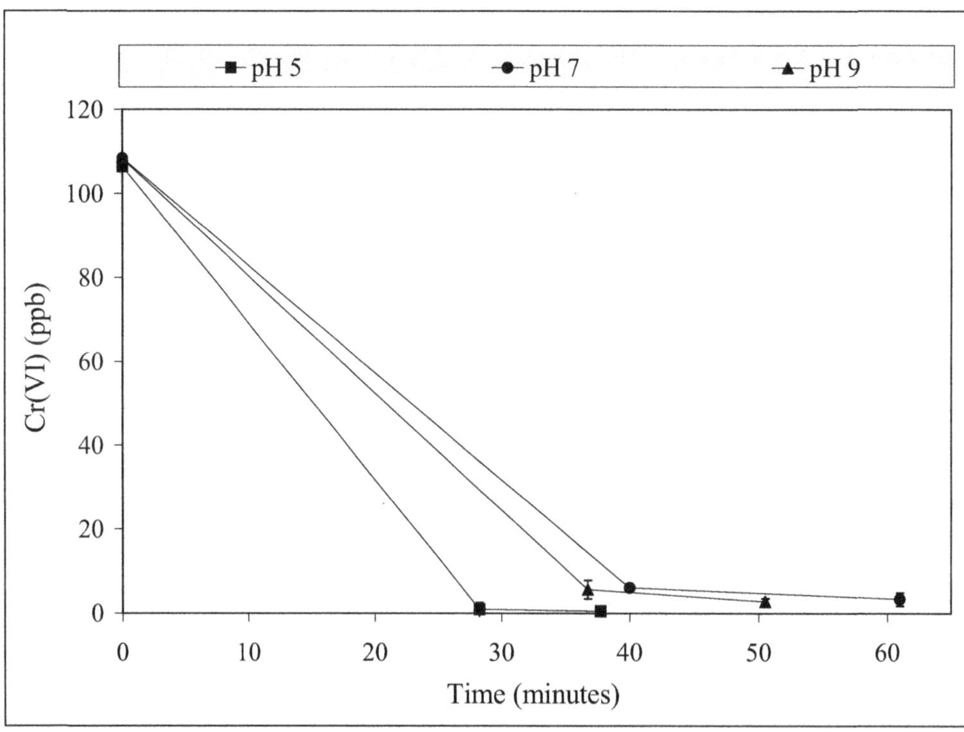

Figure 5.20 Reduction of 100 µg/L Cr(VI) by 1.3 mg/L $SnCl_2$ in deionized water (samples shown are unfiltered)

In the experiments with synthetic water at pH 7 and 9 (Figures 5.22 and 5.23), results were similar to those observed at pH 5, except that the unfiltered Cr(VI) concentrations initially decreased to nearly zero, then increased with time, and were also lower than the corresponding filtered Cr(VI) concentrations. After 200–400 min, the Cr(VI) concentrations in filtered and unfiltered samples were similar. Although it is unclear why the unfiltered samples measured lower than the filtered samples, this result could be due to interference from the particulate chromium or some form of Cr(VI) that is not measured by the colorimetric method. Whatever the cause, this is an important phenomenon because if only the unfiltered samples are measured, they would tend to overestimate the performance of $SnCl_2$ as a reductant over short times. The reason for the increase in Cr(VI) up to 200–400 min is also unclear. A parallel test with $SnCl_2$ and only 100 μg/L of Cr(III) showed that Cr(III) was not oxidized under these conditions. The subsequent increase in soluble Cr(VI) could possibly be associated with the change in precipitation or sorption kinetics in the solution. If the increase in Cr(VI) with time is real, the reaction must be allowed to proceed for a long enough time to reach equilibrium. Similar results were observed in the experiments with modified synthetic water (Figure 5.24), except that the drop in unfiltered total chromium and Cr(VI) was not as sharp—an outcome not likely attributable to the presence of particulate $CaCO_3$.

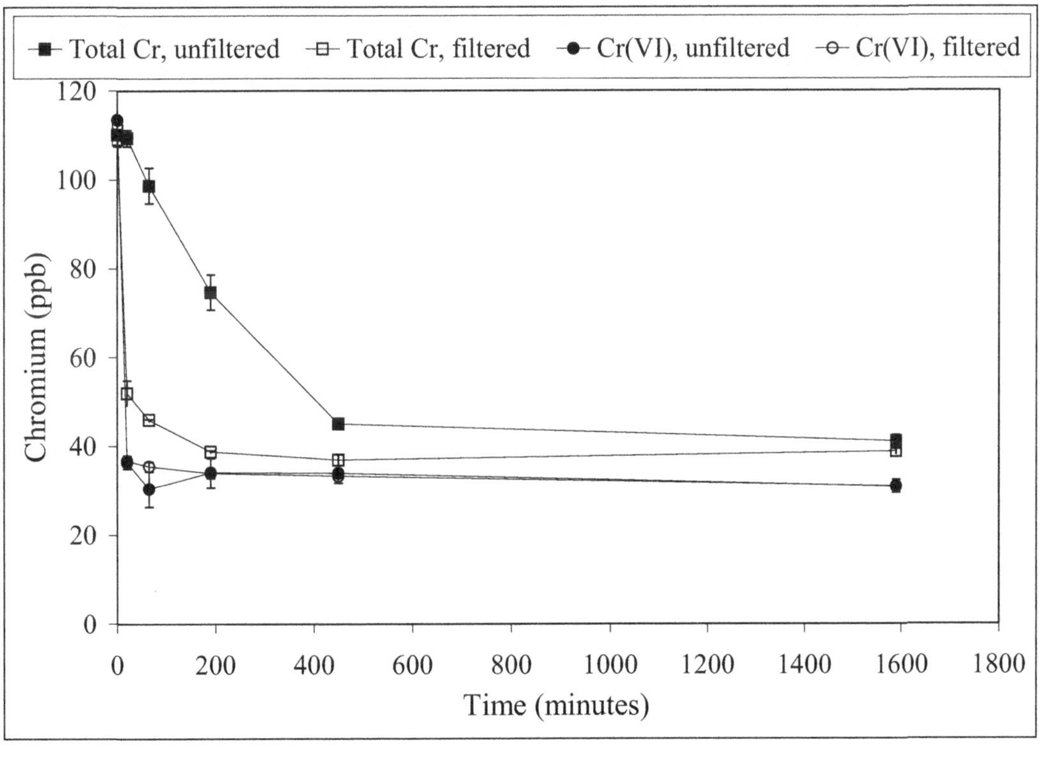

Figure 5.21 Reduction of 100 μg/L Cr(VI) by 1.3 mg/L $SnCl_2$ in synthetic water at pH 5

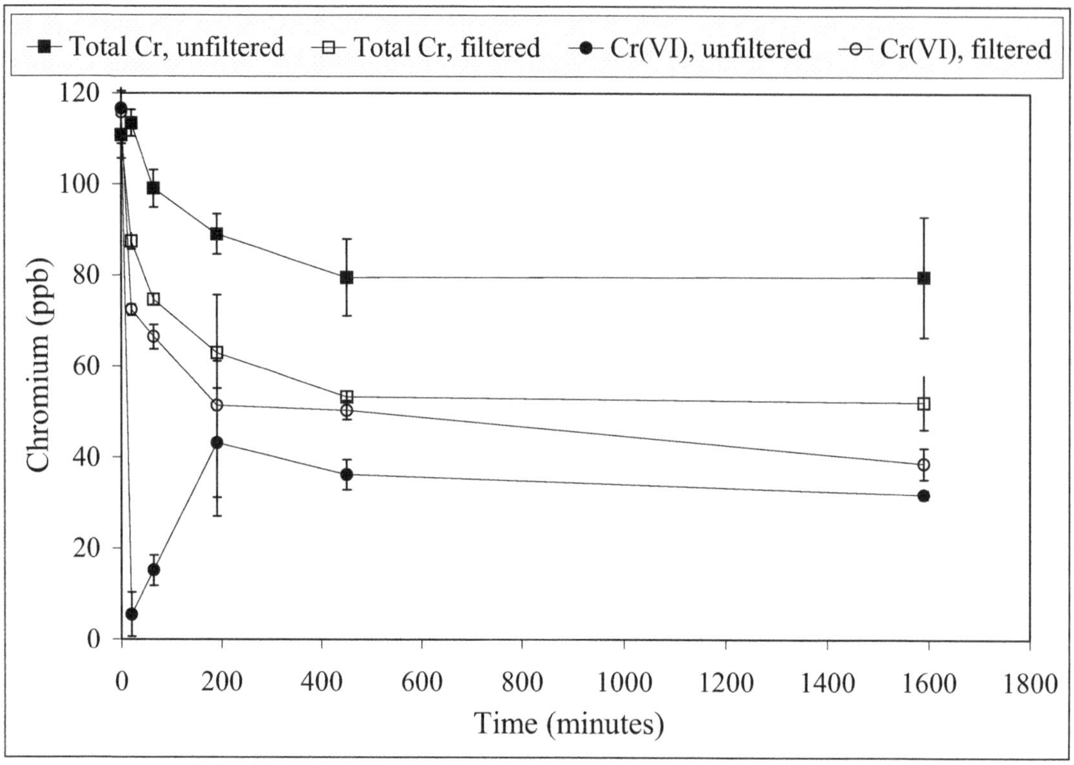

Figure 5.22 Reduction of 100 µg/L Cr(VI) by 1.3 mg/L $SnCl_2$ in synthetic water at pH 7

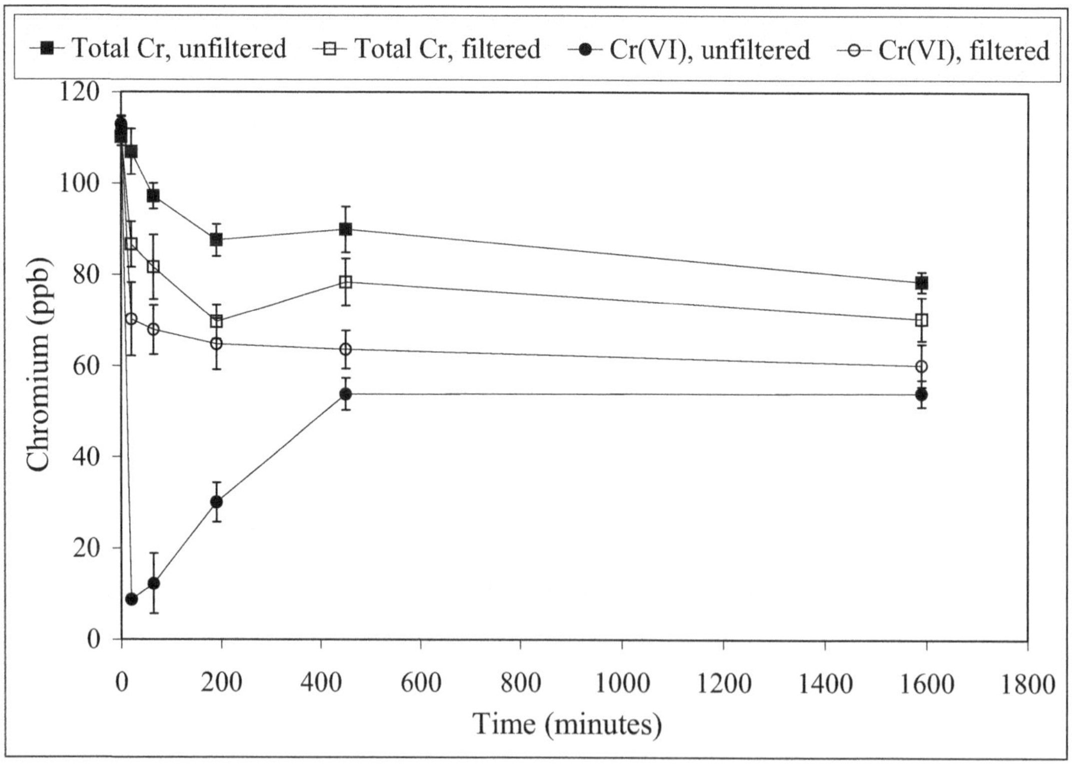

Figure 5.23 Reduction of 100 µg/L Cr(VI) by 1.3 mg/L $SnCl_2$ in synthetic water at pH 9

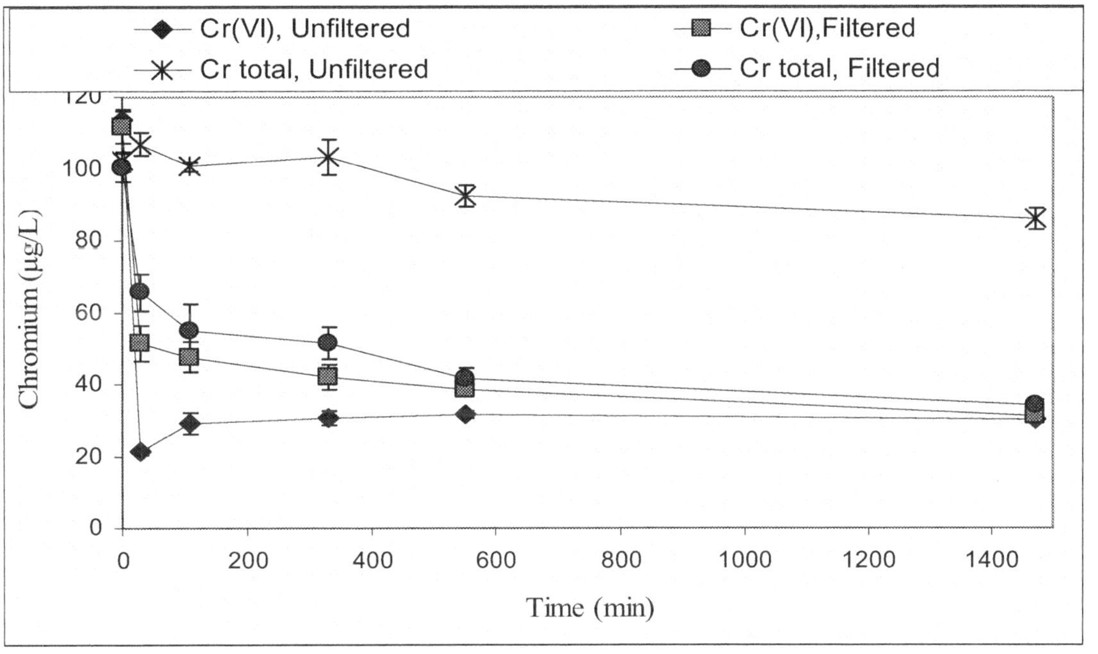

Figure 5.24 Reduction of 100 µg/L Cr(VI) by 1.3 mg/L $SnCl_2$ in modified synthetic water at pH 7

When $SnCl_2$ was tested in Glendale water at pH 5, results were similar to those observed with synthetic water; there was no difference between filtered and unfiltered Cr(VI) samples, but some chromium was precipitated (Figure B.14). At pH 7 and 9, all of the soluble chromium was Cr(VI), and the unfiltered Cr(VI) measured lower than the filtered Cr(VI) (Figures B.15 and B.16). The fact that $SnCl_2$ reduction of Cr(VI) in Glendale water was incomplete indicates that more than twice the stoichiometric dosage is needed to drive Cr(IV) reduction to completion in that water matrix.

Reduction by Sulfide

Sulfide, added as 1 mg/L S from sodium sulfide (Na_2S), was tested as a reductant for Cr(VI) at a dosage of 1 mg/L S in the DI water (Figure 5.25) and synthetic water (Figure 5.26). Reduction of Cr(VI) increased with decreasing pH, and there was less reduction at a given pH in the synthetic water than in the DI water. At all three pH conditions in both types of water, there was no difference between filtered and unfiltered samples, indicating that no precipitation occurred as the Cr(VI) was reduced to Cr(III) (data not shown).

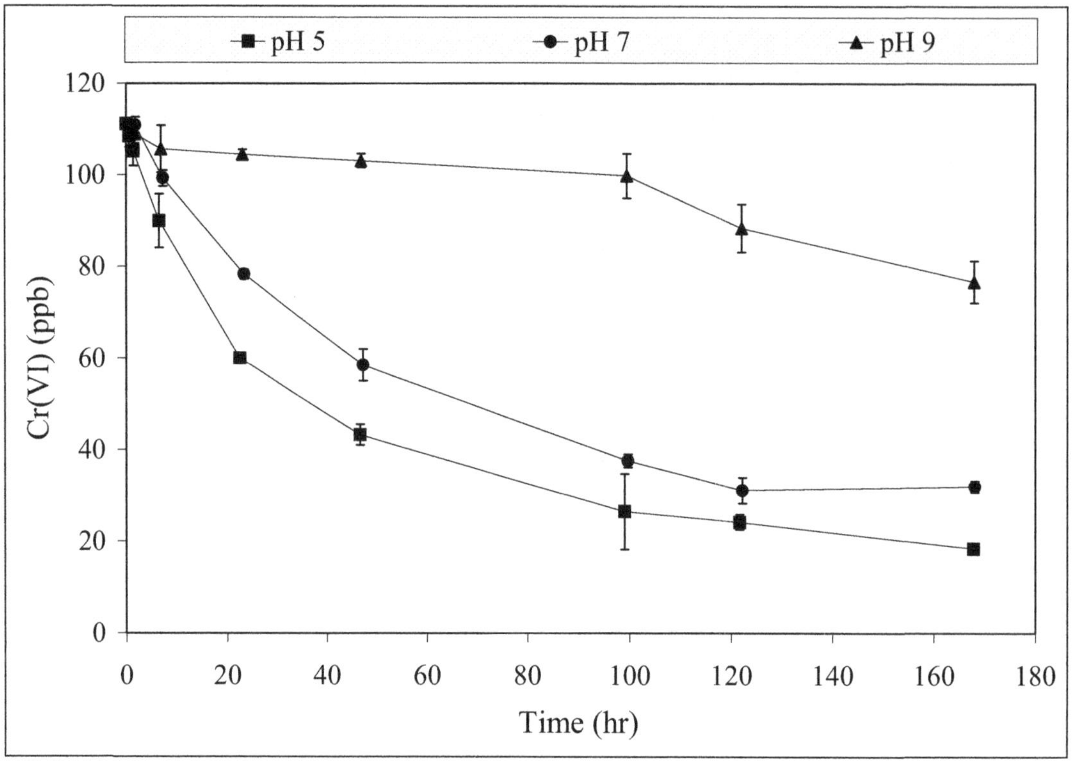

Figure 5.25 Reduction of 100 μg/L Cr(VI) by 1 mg/L sulfide in deionized water

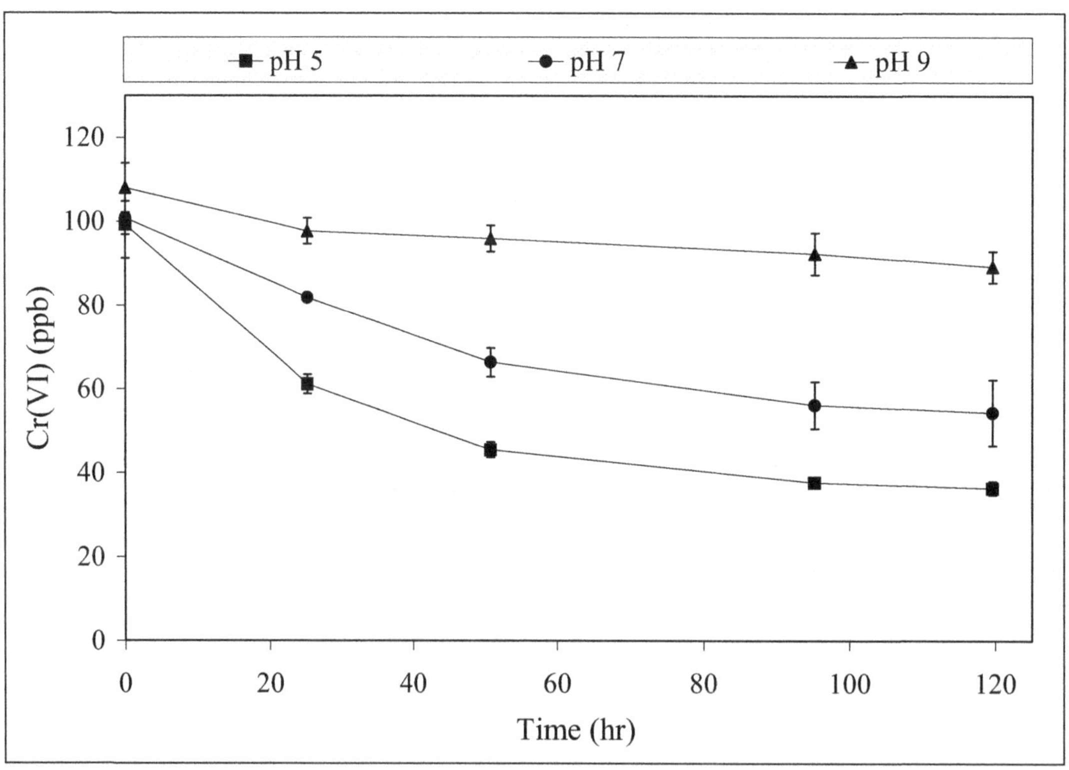

Figure 5.26 Reduction of 100 μg/L Cr(VI) by 1 mg/L sulfide in synthetic water

Reduction by Sulfite

Sulfite, added as 1 or 10 mg/L S from sodium sulfite (Na_2SO_3), was tested as a reductant for Cr(VI) in synthetic water and Glendale water. In the synthetic water, a 1-mg/L dose reduced approximately 30% of the Cr(VI), regardless of pH (Figure 5.27). The reaction had equilibrated by the first sampling time (1 hr), and no further reduction was observed over 120 hr (5 days). When the dosage was increased to 10 mg/L, the Cr(VI) reduction increased to approximately 50%, again regardless of the pH (Figure 5.28). Again, the reaction was essentially complete by the first sampling time (0.5 hr). No particulate chromium species appeared to have formed in any experiments with the synthetic water, because there was no difference between filtered and unfiltered samples (data not shown).

Because pH appeared to have little impact on the effectiveness of sulfite as a reductant, sulfite was tested only at pH 7 in the Glendale water. Although no particulate species were formed, the Cr(VI) concentration initially dropped from 100 µg/L to 35 µg/L, then increased to 80 µg/L over the following 30 hours (Figure 5.29). The reason for this subsequent increase is unclear.

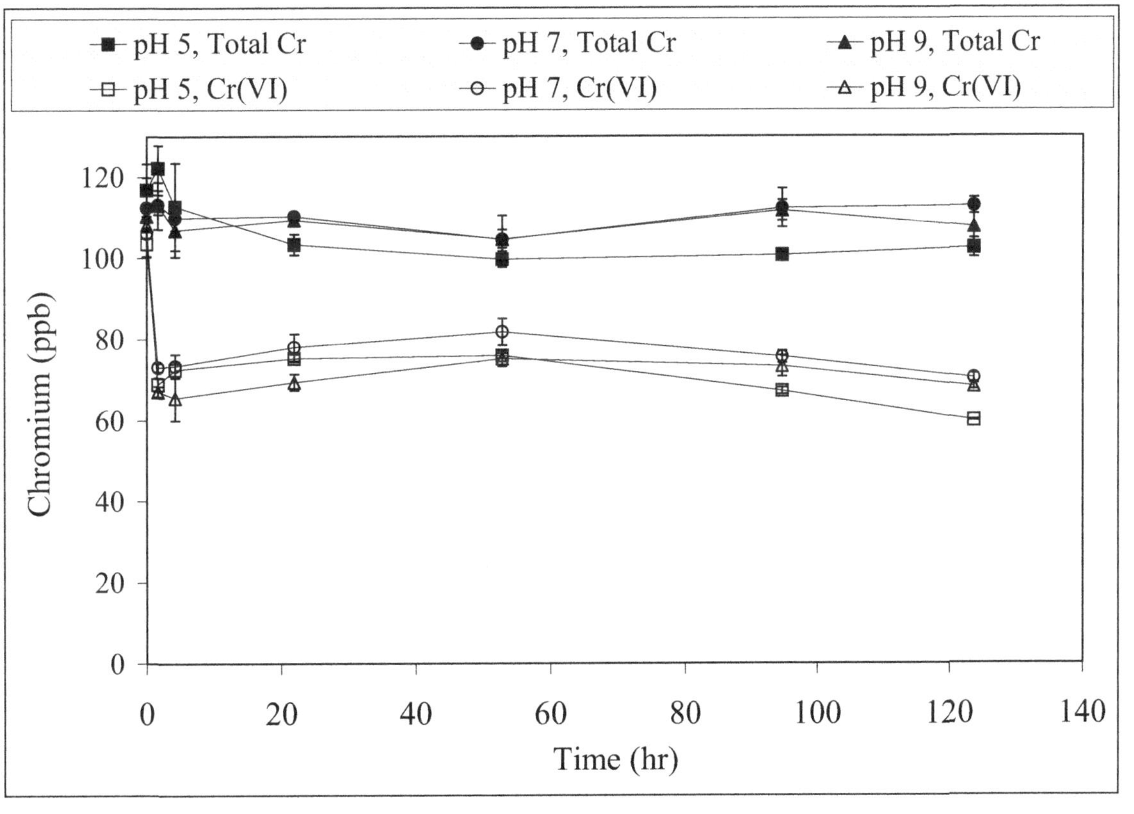

Figure 5.27 Reduction of 100 µg/L Cr(VI) by 1 mg/L sulfite in synthetic water

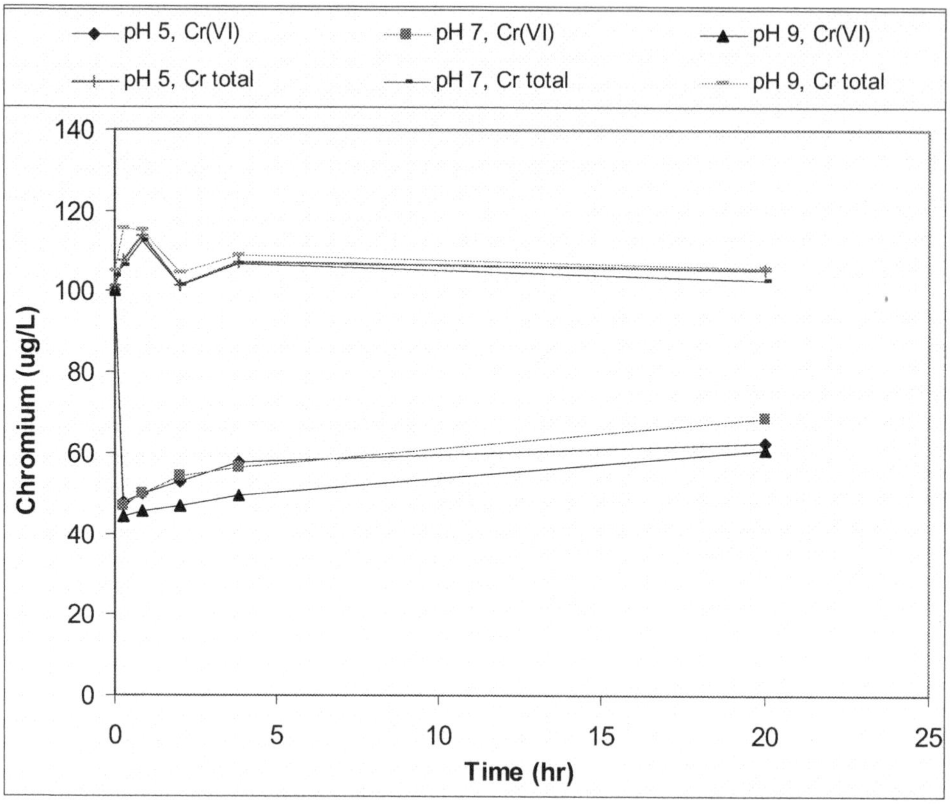

Figure 5.28 Reduction of 100 μg/L Cr(VI) by 10 mg/L sulfite in synthetic water

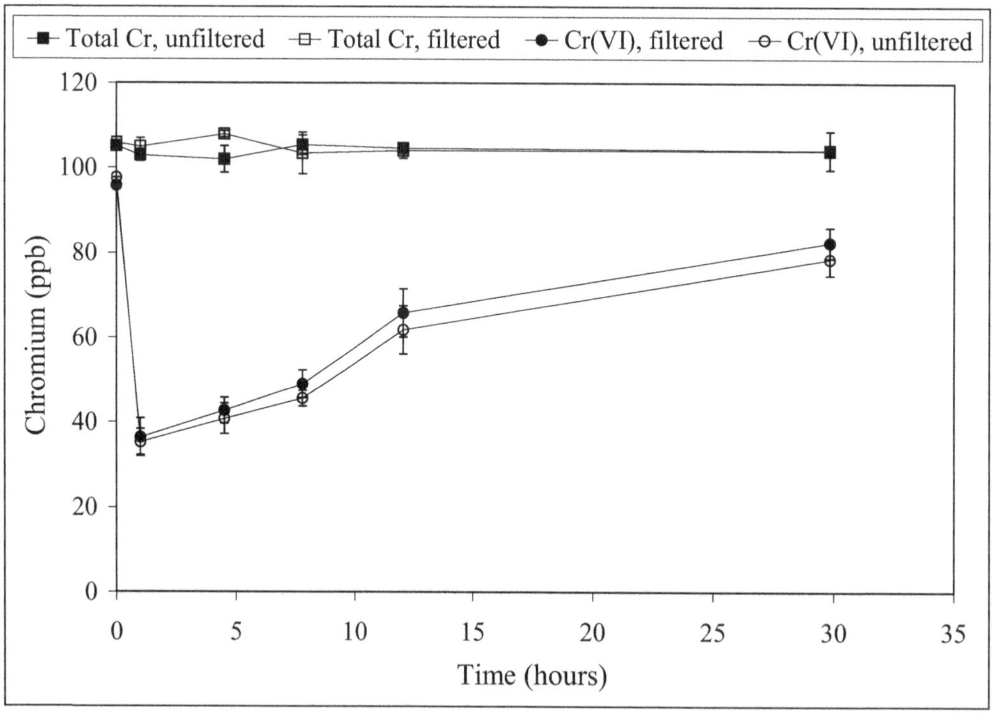

Figure 5.29 Reduction of 100 μg/L Cr(VI) by 10 mg/L sulfite in Glendale water

REOXIDATION EXPERIMENT RESULTS

Although soluble Cr(III) was shown to be oxidized by Cl_2 and KMn04, it is unknown whether particulate Cr(III) species can be oxidized. Of particular interest is the Cr(III) formed through reduction of Cr(VI) by ferrous iron (Chapter 8), which could be $Cr(OH)_{3(s)}$ or a coprecipitate mixed with Fe such as $(Cr_xFe_{1-x})(OH)_3$ (Eary and Rai 1988). These particles have been found to be small enough to pass through a 0.45-µm filter (Buerge and Hug 1997), so they may be difficult to remove from water. These reduced Cr(III) precipitates may later be exposed to oxidants in the water distribution system. Therefore, experiments with the Glendale water were conducted using permanganate and chlorine to re-oxidize the Cr(III) formed through Cr(VI) reduction by aqueous Fe^{+2} so that the percentage of Cr(III) converting back to Cr(VI) could be determined. Permanganate and chlorine were chosen on the basis of their oxidation ability in the earlier tests.

The weight ratio of Fe to Cr was 3:1, the stoichiometric dosage. Results showed that the reduction of Cr(VI) by Fe^{+2} was a very fast reaction that was complete within half an hour (the first sampling time). Only a 30–40% reduction of the Cr(VI) was observed, and no further reduction occurred even after 10 days' reaction (Figure 5.30). This incomplete reduction is likely due to the need for a higher dosage of Fe^{+2}, competition for the Fe^{+2} from dissolved oxygen or other species in the Glendale water, or, possibly, the low reaction temperature (15°C). The unfiltered total chromium dropped from 100 µg/L to around 70 µg/L, and the unfiltered total chromium concentration tracked unfiltered Cr(VI) after about 100 hr. This result might have been caused by some loss of Cr(III) during the solution transfer, but acid digestion of the containers used to transfer the solution indicated that no loss of Cr(III) species had occurred as a result of sorption to the container walls. Thus, it appears that the loss of total chromium was due to the particles not being sampled. All the filtered Cr(VI) and total chromium concentrations were essentially the same (Figure 5.30), indicating that as soon as the Cr(VI) was reduced to Cr(III), a particulate species was formed and removed by the filter. However, there was no difference between the 0.45-µm and 0.2-µm filtered samples (data not shown).

Chlorine Reoxidation Experiment

After 96 hours of reaction time, 1 mg/L of chlorine was added to one aliquot of the previous Fe^{2+} test solution to investigate the chlorine's ability to re-oxidize the reduced Cr(III) species. Results showed that essentially no reduced Cr(III) was re-oxidized to Cr(VI) after 6 days (Figure 5.31). No chlorine residual measurements were taken during this experiment.

Permanganate Reoxidation Experiment

After 96 hours of reaction time, 3 mg/L of $KMnO_4$ was added to one aliquot of the previous Fe^{2+} test solution to investigate the permanganate's ability to re-oxidize the reduced Cr(III) species. Results were similar to those from the chlorine test; no oxidation of Cr(III) to Cr(VI) had occurred after about 250 min (Figure 5.32). Because the time frame for the $KMnO_4$ reoxidation test is much shorter (250 min) than the previous reduction step (5 days, see Figure 5.30), Figure 5.32 shows only the reoxidation portion of the experiment.

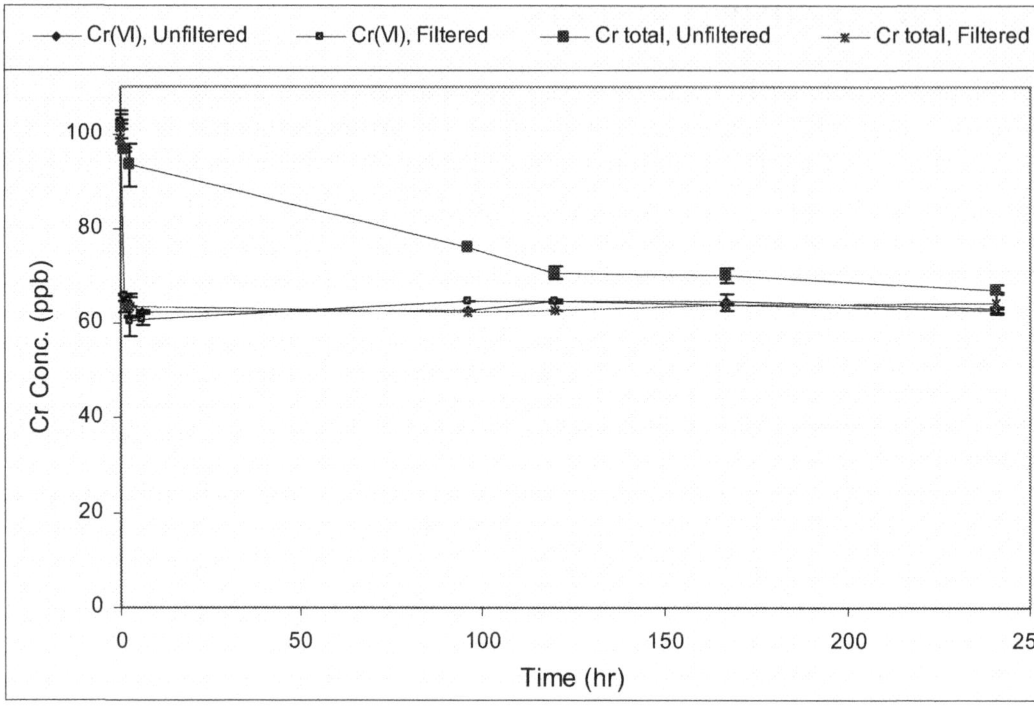

Figure 5.30 Control test for reoxidation experiment—reduction of 100 μg/L Cr(VI) by ferrous iron in Glendale water at pH 7

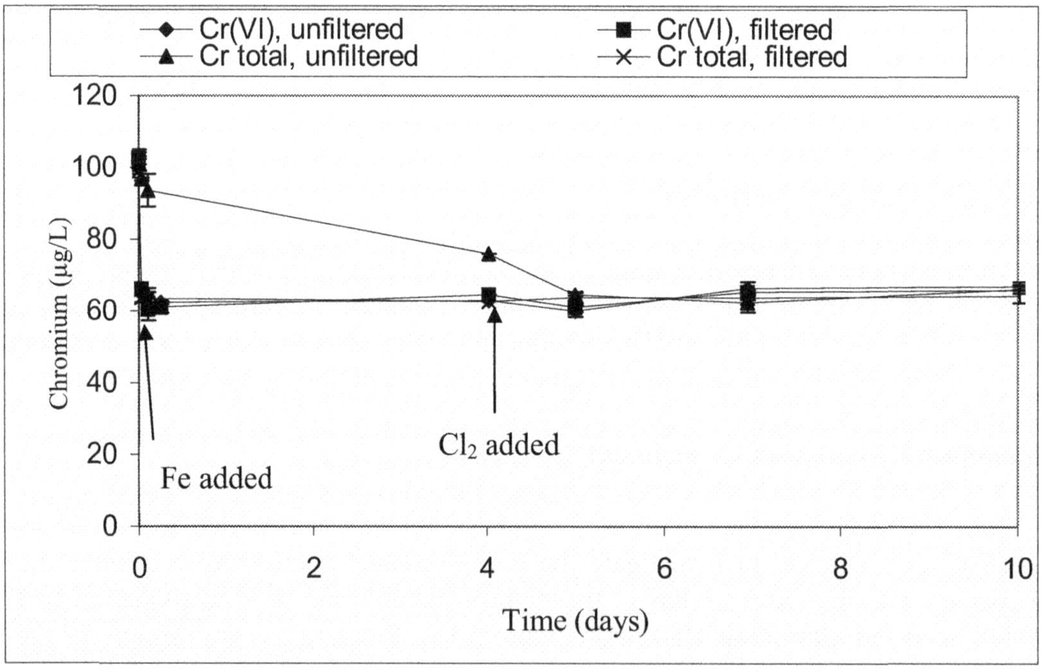

Figure 5.31 Reduction of 100 μg/L Cr(VI), followed by reoxidation by 1 mg/L chlorine

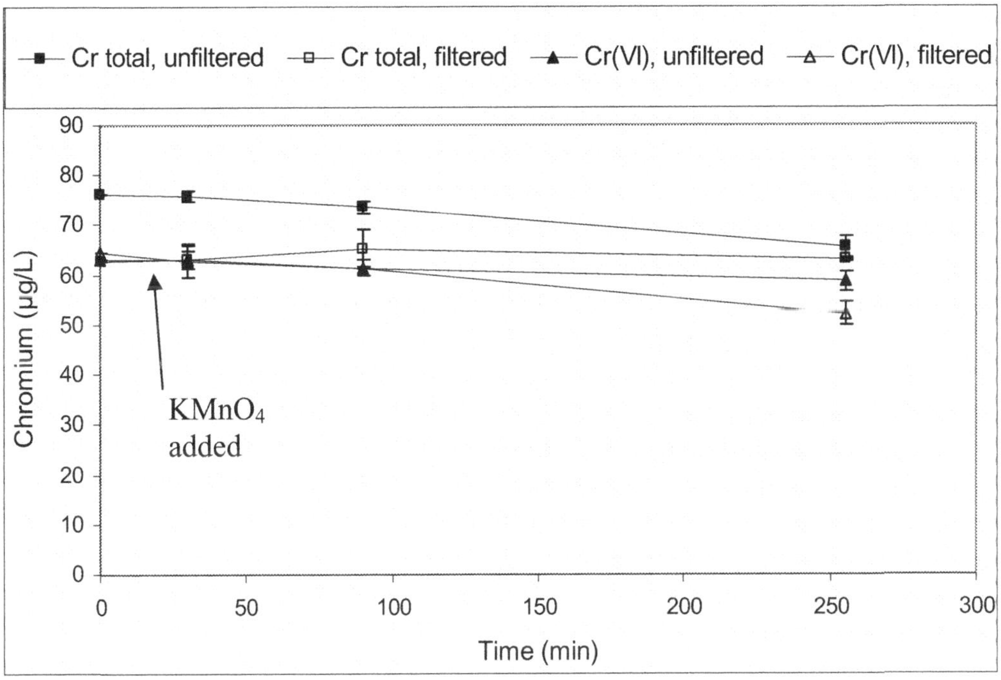

Figure 5.32 Reoxidation of 100 μg/L Cr(VI) by 3 mg/L permanganate

ADDITIONAL HEXAVALENT CHROMIUM REDUCTION AND CHLORAMINE REOXIDATION RESULTS

In initial experiments described in this report, little to no oxidation of Cr(III) was observed when a low concentration (0.5 mg/L) of chloramines was added (Figures 5.12 and 5.13). This finding contrasted with earlier bench-scale work that showed chloramines could slightly reoxidize Cr(III) to Cr(VI) (McGuire Environmental Consultants 2002). Additional experiments were conducted to more explicitly address whether chloramines could reoxidize stannous-reduced Cr(VI) in Glendale water.

Several changes in the experimental methods were instituted so that chloramine oxidation effects could be examined more closely. In this work, Cr(VI) concentrations were measured by ion chromatography (IC), which offers a significant advantage over the colorimetric method used in the initial experiments. Second, the Cr(VI) was only partially reduced to Cr(III) to allow the observation of changes well above detection limits and to exhaust the reduction potential of stannous chloride (i.e., to prevent chloramine reduction by excess stannous chloride). Finally, the chloramine concentrations were monitored over the time period of the experiments.

First, the 100 ug/L Cr(VI) that had been added to Glendale water was reduced to Cr(III) by approximately 70% using a stannous chloride stock solution of 309 mg/L. The Cr(VI) reduction was achieved with a stannous chloride dose twice that of the stoichiometric dosage (2 mL of stock added to 1 L), yielding Cr(VI) concentrations of 25–30 μg/L. This dosage was selected to achieve incomplete reduction, thereby exhausting the stannous chloride's reduction potential for the chloramines.

In separate bottles, five different concentrations of chloramines with a 3:1 mass ratio of Cl_2 to NH_3-N were added to the water containing partially reduced chromium. A control bottle with no chloramines added was also carried through the experiments. Ammonia was added to the bottles first to prevent oxidation of Cr(III) by free chlorine. Figure 5.33 shows the five chloramine concentrations (plus the control) used and the changes in concentration that occurred over the course of the experiment. All experiments were conducted at an average pH of 7.0 (adjusted from an average pH of 7.3 in the Glendale water using dilute HCl).

The bottles containing the mixed Cr(III) and Cr(VI) species and chloramine concentrations were shipped to USU for Cr(VI) analysis by IC. Subsamples were measured for total chlorine residual and Cr(VI) at 48 hr and 168 hr, periods representing reasonable detention times in water utility distribution systems.

Figure 5.34 shows that the addition of stannous chloride yielded a Cr(VI) concentration of 28 µg/L. At 48 hr, Cr(VI) had increased to 42–51 µg/L. Negligible change was observed between 48 and 168 hr. The percentage of Cr(III) reoxidation was approximately 50–82%, significantly higher than that observed in the initial experiments in this study. Therefore, the presence of even low concentrations of chloramines can oxidize significant concentrations of Cr(III) to Cr(VI) over several days. If a utility wishes to control Cr(VI) by reducing all of the Cr(VI) in its source water to the less toxic form of Cr(III), this research shows that Cr(VI) control solely by reduction is not a viable option.

SUMMARY OF FINDINGS

The oxidation of Cr(III) to Cr(VI) and the reduction of Cr(VI) to Cr(III) were investigated under five water quality conditions: DI water with 10^{-3}M $NaNO_3$, synthetic water, modified synthetic water without $CaCO_3$ precipitate, reducing water, and Glendale water. Four oxidants were tested: dissolved oxygen, free chlorine, chloramine, and potassium permanganate. Three reductants were tested: stannous chloride, sulfide, and sulfite. In all the water quality conditions, Cr(VI) was stable if no reductants were added. Cr(III) was generally stable at pH 5 and 7, but at pH 9, 70–80% of the Cr(III) formed a particulate species that was removed by a 0.45-µm filter. The formation of Cr(III) precipitates was also observed when initial chromium concentrations of 10, 20, and 40 µg/L were used. The presence of $CaCO_3$ solids was not responsible for the formation of this particulate Cr(III) species.

Oxidation by Dissolved Oxygen

Dissolved oxygen present in water at saturation was not able to oxidize any of the Cr(III). Thus, in the absence of any other oxidants, Cr(III) can be expected to remain stable if it does not precipitate.

Oxidation by Free Chlorine

A 1-mg/L dose of Cl_2 oxidized some Cr(III) to Cr(VI). A maximum of 90% oxidation was observed in the synthetic water after 140 hr (5.8 days). Oxidation was approximately 20% lower in the Glendale water at each pH. In the reducing water, which contained dissolved manganese and NOM, no oxidation at all was observed. Although less than 10% oxidation was observed in any of the water samples tested at pH 9, this result could be due to the fact that much of the added Cr(III) was in particulate form. Some of the oxidation effects could be explained by the protocol used for the free chlorine oxidation experiments because the protocol did not include measuring initial chlorine residuals or following chlorine degradation over time.

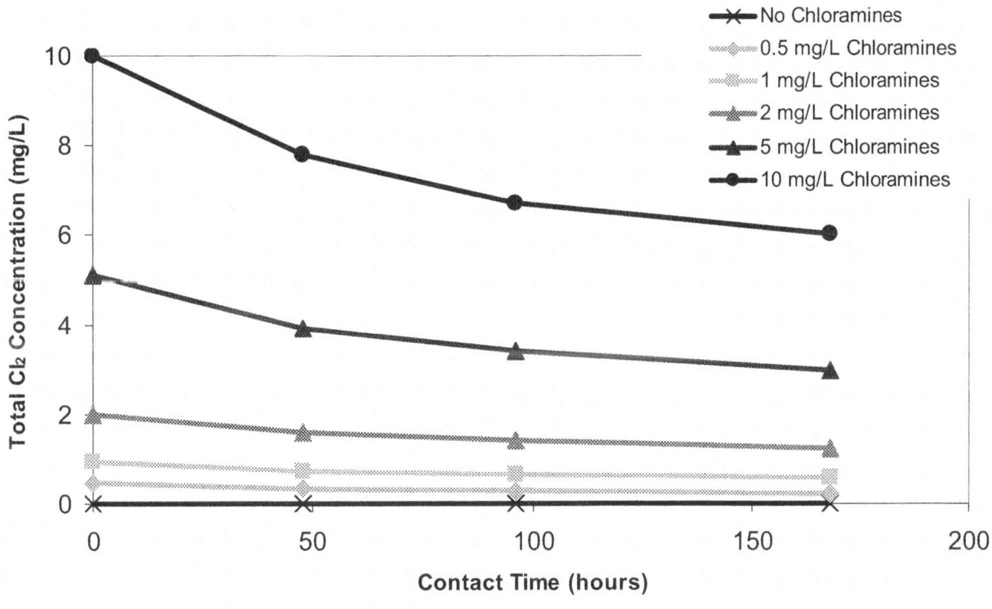

Figure 5.33 Combined chlorine (chloramine) concentrations over the experimental period

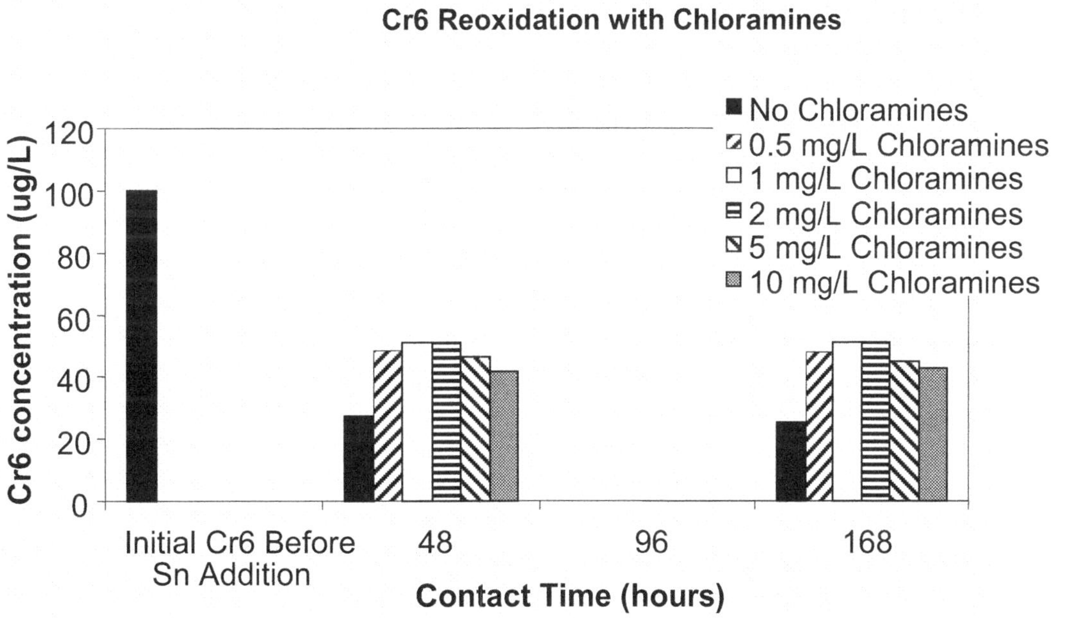

Figure 5.34 Cr(III) reoxidation to Cr(VI) with different chloramine doses

Considering the behavior of chromium released into a water distribution system, water containing a free chlorine residual may oxidize soluble Cr(III) to Cr(VI). Particulate Cr(III) species formed through reduction of Cr(VI) by ferrous iron would not likely be oxidized by Cl_2. Given the potential for oxidation of soluble Cr(III) in the distribution system, simply removing the Cr(VI) from water in anticipation of meeting a Cr(VI) MCL, without addressing Cr(III) removal, will not be a viable option.

Oxidation by Chloramines

Findings during initial experiments showed that oxidation of Cr(III) by chloramine (dose of 0.5 mg/L as Cl_2) was negligible at the pH conditions studied in both synthetic and deionized water. Additional tests showed that chloramine concentrations typical of those found in distribution systems oxidized significant amounts (up to 23 μg/L) of Cr(III) to Cr(VI) in reasonable periods of time. Therefore, simply reducing Cr(VI) to Cr(III) will not control Cr(VI) in a distribution system containing chloramines.

Oxidation by Permanganate

Potassium permanganate was tested as an oxidant for Cr(III) under all five water quality conditions. In the DI water at pH 7 and 9, oxidation was essentially complete within 1 hr but required 24 hours for complete oxidation at pH 5. In the synthetic water, a similar trend was observed, although less time (about six hr) was required for complete oxidation at pH 5. Consistent with the control, Cl_2, and chloramine experiments, nearly 60% of the chromium was in particulate form at pH 9. However, a significant difference was that permanganate appeared to be able to oxidize this particulate Cr(III). There was no linear relationship between Cr(III) oxidation and permanganate dosage in the range of 1–10 mg/L as $KMnO_4$, and a 3-mg/L dose was apparently the most effective for the water quality conditions tested in this study.

If a treatment scenario included converting all Cr(III) in a water supply to Cr(VI) prior to anion exchange, permanganate might be an effective oxidant, although determining the optimum dosage would require testing with the specific water quality conditions. Water quality and pH will have a significant impact on the effectiveness of permanganate. Because $KMnO_4$ reacts with the diphenylcarbohydrazide reagent used in the colorimetric determination of Cr(VI), future researchers should be cautious when interpreting results.

Oxidation by Hydrogen Peroxide

Hydrogen peroxide could not be evaluated as an oxidant for Cr(III) in this study because the H_2O_2 also oxidized the colorimetric reagent used to measure Cr(VI). However, because peroxide is listed as a Standard Method, it is likely to be an effective pretreatment chemical for Cr(III) oxidation.

Reduction by Stannous Chloride

Stannous chloride was tested at 1.3 mg/L (twice the stoichiometric dosage) as a reductant for Cr(VI). In the DI water, $SnCl_2$ reduced all of the Cr(VI) within 30 min, whereas in the synthetic water, it reduced only 40–60% of the Cr(VI). Reduction was slightly better at lower pH, but overall the reaction did not appear to be very sensitive to pH. More complete reduction could have been accomplished with stannous chloride in the synthetic water if a higher multiple of the stoichiometric dosage had been used.

As Cr(VI) was reduced to Cr(III), some of the Cr(III) formed a particulate species. Moreover, at pH 7 and 9, Cr(VI) concentrations in the unfiltered samples initially appeared to drop below detection limits but then increased with time. These unfiltered Cr(VI) concentrations also appeared to be lower than those in the filtered samples, which is not physically possible. Similar results were obtained in the experiments with Glendale water. It appears that there is some kind of interference with the measurement of Cr(VI), possibly from the particulate species that forms as Cr(VI) is reduced to Cr(III). This is important because if the full amount of Cr(VI) is not being measured in unfiltered samples, the effectiveness of $SnCl_2$ may be overestimated.

Reduction by Sulfide

Sulfide is not typically added to drinking water, but it may be present in some reduced groundwater supplies, so its effectiveness as a reductant was evaluated in DI and synthetic water. Sulfide reduction of Cr(VI) in DI water increased with decreasing pH, with a maximum of 80% reduction at pH 5. In the synthetic water, approximately 20–30% less reduction occurred at a given pH compared with the DI water. Although the Cr(VI) was being reduced to Cr(III), no particulate species were formed at any of the three pH conditions tested, an outcome that contrasted with the results of the $SnCl_2$ experiments.

Although sulfide is a relatively effective reductant for Cr(VI), it requires a time frame of several days. A water treatment plant would be unlikely to add sulfide as a treatment chemical, but water supplies containing reduced sulfur may contain Cr(III) instead of Cr(VI).

Reduction by Sulfite

Sulfite, added as sodium sulfite, was tested in the synthetic water and the Glendale water. At a dose of 1 mg/L as S, the sulfite reduced 30% of the Cr(VI), whereas a dose of 10 mg/L as S reduced 50% of the Cr(VI), within 1 hr. The pH had no effect on the effectiveness of the sulfite under any conditions tested. As with sulfide, no particulate species were formed when the Cr(VI) was reduced to Cr(III). An unusual result was observed in the Glendale water—the Cr(VI) concentration initially decreased, then increased, although no particulate chromium was formed.

Sulfite may be an effective reductant, but its dosage will likely have to be increased significantly above 10 mg/L for the conditions studied in these experiments.

CHAPTER 6
SORPTION AND ANION EXCHANGE TECHNOLOGY ASSESSMENT

BATCH EXPERIMENTS

The batch experiments served as a screening mechanism for determining which media removed Cr(VI) from water and which did not. The batch adsorption experiments consisted of the following individual experiments:

- Initial capacity experiment,
- Refined carbonate-buffered capacity experiment,
- pH sensitivity experiment, and
- Dual solute experiments.

The methods used to conduct the batch experiments are described in Chapter 3. The media tested in these studies are identified in Table 3.3.

Initial Capacity Experiment

The initial capacity experiment was conducted to provide a preliminary idea of treatment media performance. Figure 6.1 displays the 24-hr residual Cr(VI) concentrations from the initial capacity experiment. With the exception of media D, all the treatment media tested exhibited some degree of ability to remove Cr(VI). As shown in Figure 6.1, many of the 24-hr Cr(VI) concentrations from these media doses were at or near the Hach method detection limit of 7 μg/L.

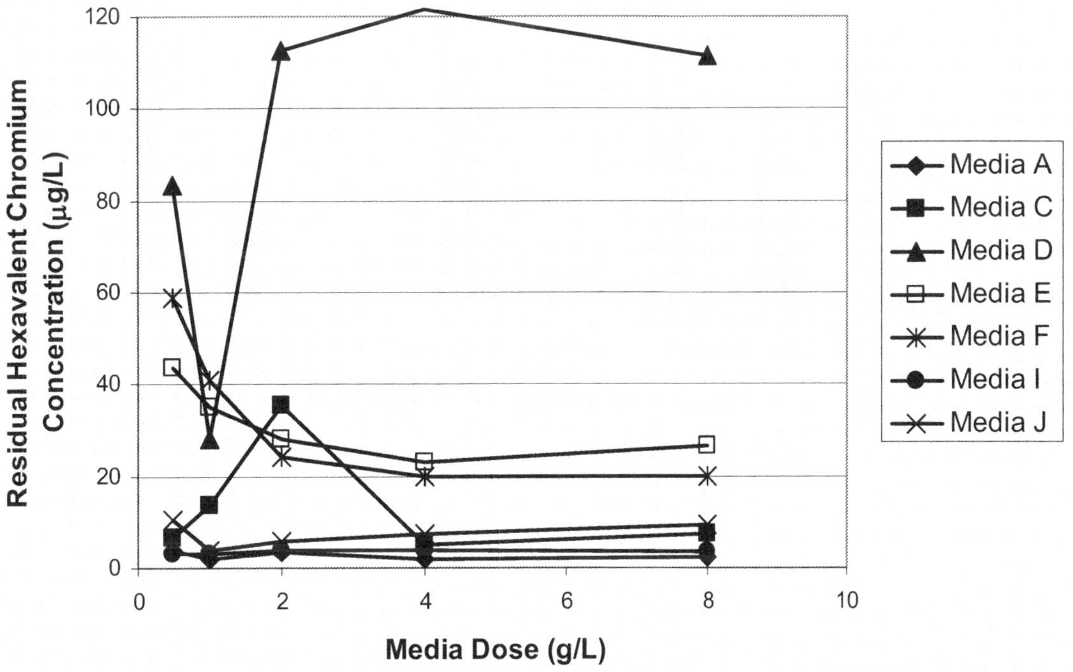

Figure 6.1 Results of initial capacity experiment—24-hr samples

Refined Carbonate-Buffered Capacity Experiment

Following the initial capacity experiment, the refined carbonate-buffered capacity experiment was developed to control pH during the batch experiments. In this experiment, the media dosages were also reduced to yield more detectable residual Cr(VI) concentrations. From the experimental matrix of five media doses sampled at three different times, the 1-g/L media dose sampled at 2 hr provides a good picture of the capacities of the individual media. Figure 6.2 shows the Cr(VI) removal efficiency of each treatment media evaluated in the refined carbonate-buffered capacity experiment.

Media C and I showed excellent performance, with more than 90% removal of Cr(VI). Media J showed good performance, with Cr(VI) removal of 50–90%. Media A, B, D, E, F, and G showed poor performance, with less than 50% removal. Media H showed no ability to remove Cr(VI) in the initial capacity experiment or the refined carbonate-buffered capacity experiment.

Treatment media performance varied as a function of media dose and sampling time. Figures 6.3 and 6.4 show residual Cr(VI) concentrations achieved by various treatment media as a function of media dose and sampling time, respectively.

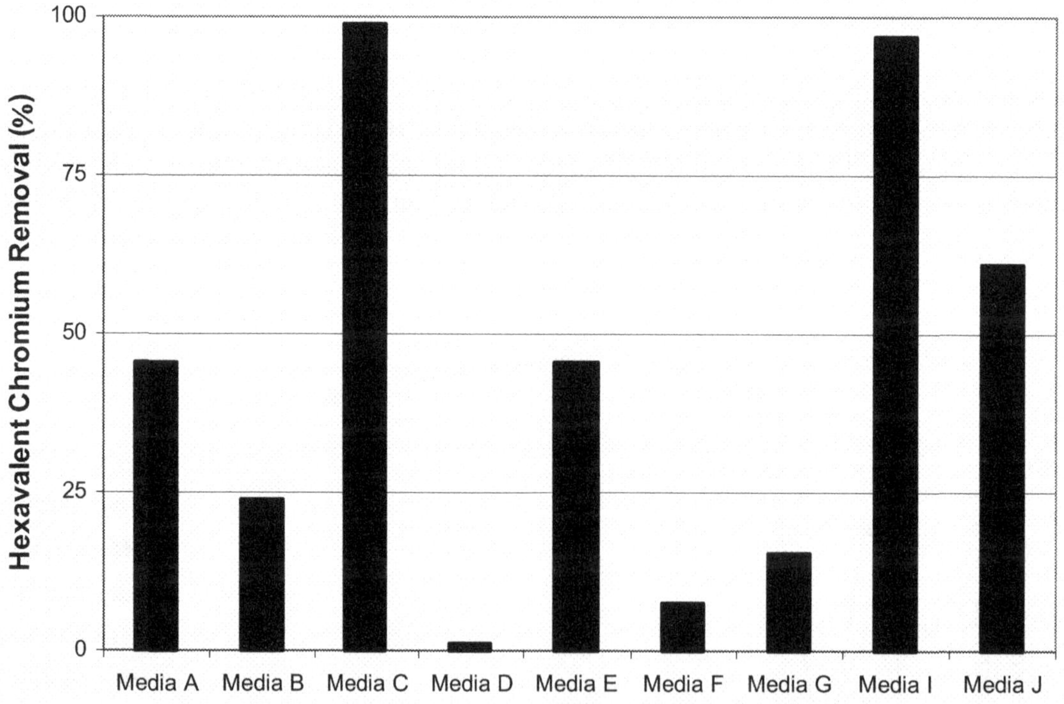

Figure 6.2 Results of refined carbonate-buffered capacity experiment using a 1-g/L media dose—2-hr samples

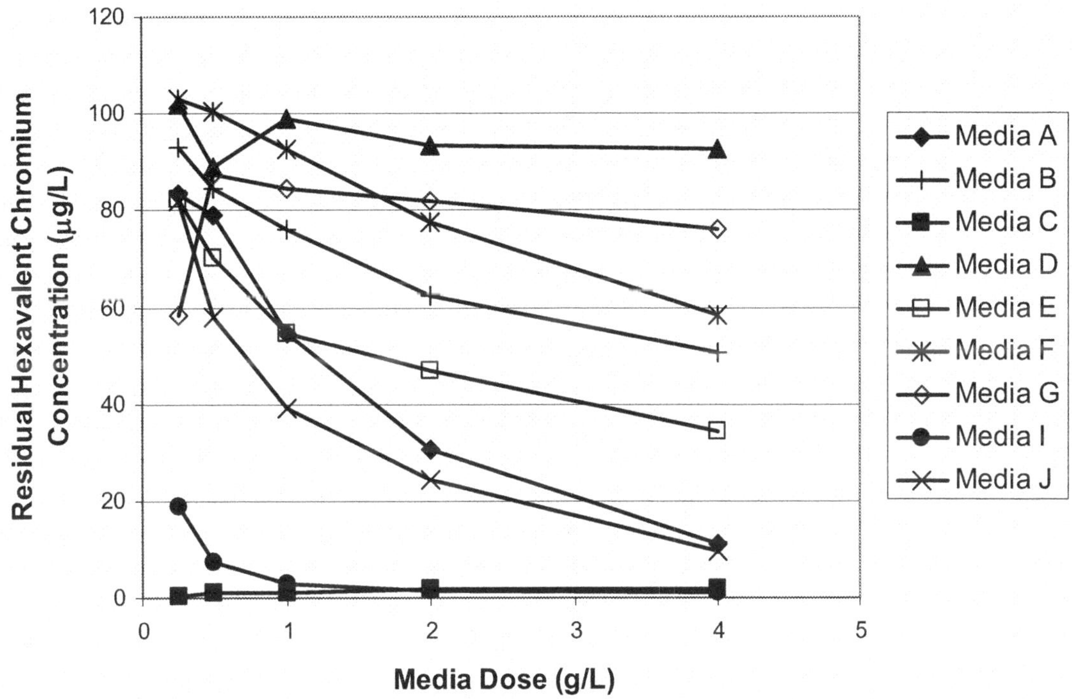

Figure 6.3 Increased Cr(VI) removal with increased media doses after 2 hours

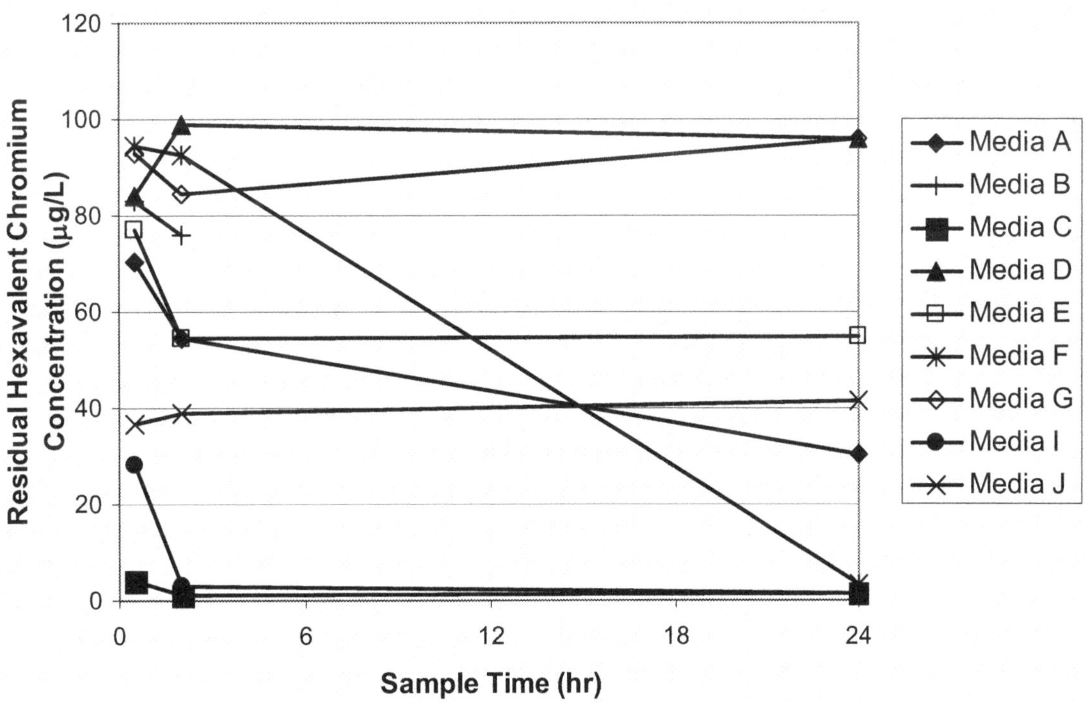

Figure 6.4 Effect of time on Cr(VI) removal in experiments using a 1-g/L media dose

Several treatment media that showed poor performance at removing Cr(VI) at the 2-hr sampling time based on the 1-g/L media dose showed improved performance with increasing dosages (Figure 6.3). Media A strongly exhibited this effect. On the other hand, increasing dosages of media D and G showed little to no improvement at removing Cr(VI).

The kinetics of Cr(VI) removal was investigated by measuring residual Cr(VI) concentrations as a function of sampling time (in this case, with a 1-g/L media dose). Media F exhibited a strong kinetic effect, improving Cr(VI) removal by more than 50% from 30 min to 24 hr. Media A, B, E, and I exhibited poor improvement in Cr(VI) removal from 30 min to 24 hr. Media C, D, G, H, and J showed no improvement in Cr(VI) removal from 30 min to 24 hr.

Table 6.1 summarizes the results of the refined carbonate-buffered capacity experiment in terms of both Cr(VI) removal and the effect of time.

Media C and I performed best at removing Cr(VI), and media J showed good removal performance. The remaining media exhibited poor removal performance; however, media F showed excellent performance at 24 hr. Media D and H were not used in subsequent experiments because they did not exhibit any capability for removing Cr(VI) in either the initial capacity experiment or the refined carbonate-buffered capacity experiment.

Table 6.1
Summary of results from refined carbonate-buffered capacity experiment

Media identification	Cr(VI) removal performance*	Kinetic effect†
Media A	Poor	Weak
Media B	Poor	Weak
Media C	Excellent	None
Media D	Poor	None
Media E	Poor	Weak
Media F	Poor	Strong
Media G	Poor	None
Media H	Poor	None
Media I	Excellent	Weak
Media J	Good	None

* Cr(VI) removal performance based on a 2-hr equilibrium concentration with a 1-g/L media dose

- Excellent = >90% removal
- Good = 50–90% removal
- Poor = <50% removal

† Kinetic effect shows improved $Cr(VI)^{+}$ removal from 30 min to 2 hr and 24 hr

- Strong = >50% improvement in removal
- Weak = 10–50% improvement in removal
- None = <109% improvement in removal

pH Sensitivity Experiment

The pH sensitivity experiment was conducted after the initial and refined capacity experiments to evaluate the effect of pH on Cr(VI) removal by sorptive media. Figure 6.5 displays the pH sensitivity results for the experiment using a 1-g/L media dose and sampled at 2 hr.

The treatment media affected by pH typically showed improved performance at removing Cr(VI) in lower pH conditions. Media G differed from this trend in that it performed best in neutral pH conditions. Media C and I were effective in all pH conditions under these experimental conditions. Table 6.2 shows which pH condition was associated with each treatment media's best removal performance in the pH sensitivity experiment. The extent of the pH effect is also characterized.

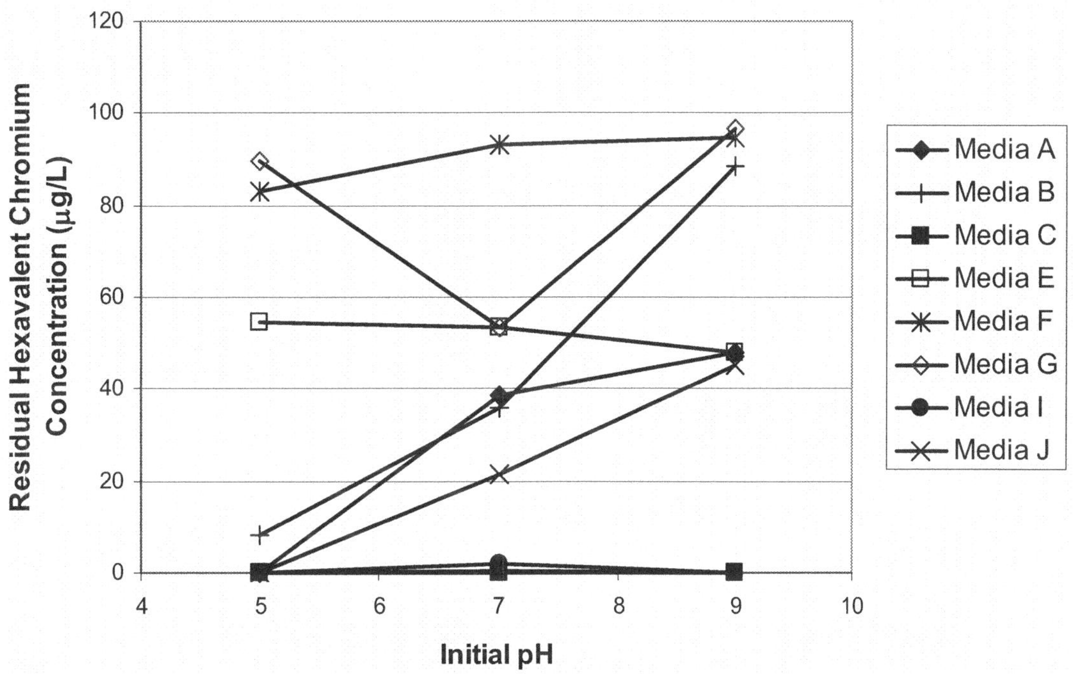

Figure 6.5 Results of pH sensitivity experiment using a 1-g/L media dose—2-hr samples

Table 6.2
Results of pH sensitivity experiment

Media identification	pH effect
Media A	Acidic pH preferred; strong effect
Media B	Acidic pH preferred; strong effect
Media C	Excellent removal at all pH conditions; no effect
Media E	Moderate removal at all pH conditions; no effect
Media F	Acidic pH preferred; weak effect
Media G	Neutral pH preferred; strong effect
Media I	Excellent removal at all pH conditions; no effect
Media J	Acidic pH preferred; strong effect

Because of their moderate Cr(VI) removal performances, media E and G were not used in subsequent experiments. Media F was carried forward to subsequent experiments on the basis of its excellent Cr(VI) removal performance at the 24-hr sampling time.

Dual Solute Experiments

Dual solute experiments were conducted to determine the impact of competing solutes on Cr(VI) removal. Co-removal of the dual solutes was also evaluated. The following dual solutes were evaluated at three concentrations, as detailed in Chapter 3:

- Chloride
- Sulfate
- Bicarbonate
- Silicate
- TOC
- Phosphate
- Nitrate
- Fluoride
- Arsenic
- Perchlorate

In these experiments, media C (sulfur-modified iron) was evaluated at a 0.1-g/L dose rather than the 1.0-g/L dose used for all other media because of its much higher capacity for removing Cr(VI).

Competitive Effects of Dual Solutes

Of all the dual solutes, sulfate and bicarbonate negatively affected Cr(VI) removal to the greatest extent. Figures 6.6 and 6.7 show the competitive effects of sulfate and bicarbonate on Cr(VI) removal.

The Cr(VI) removal performance of all treatment media diminished in the presence of increasing sulfate. In the presence of bicarbonate, all treatment media except media C and I exhibited diminished Cr(VI) removal performance.

Table 6.3 characterizes the dual solute effects on Cr(VI) removal by all treatment media evaluated in the dual solute experiments. The table shows which treatment media were affected by multiple dual solutes and which dual solutes affected multiple treatment media. For example, media C stands out as a treatment media that was basically unaffected by the dual solutes, whereas media A and F were affected by several dual solutes. From the other perspective, sulfate, bicarbonate, and TOC diminished the removal performance of multiple treatment media.

Co-removal of Dual Solutes

In addition to identifying which dual solutes diminished Cr(VI) removal performance, these experiments identified which treatment media could also remove co-occurring contaminants such as arsenic, perchlorate, and nitrate. Table 6.4 shows the percent removal of arsenic, perchlorate, and nitrate averaged from the three initial dual solute concentrations.

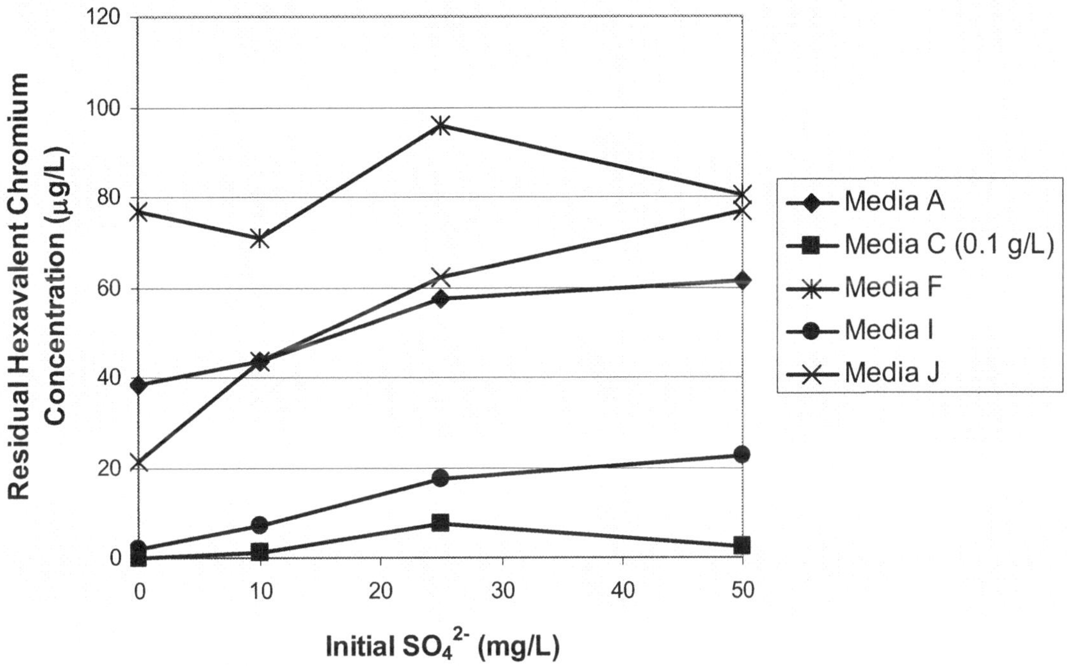

Figure 6.6 Effect of sulfate on hexavalent chromium removal

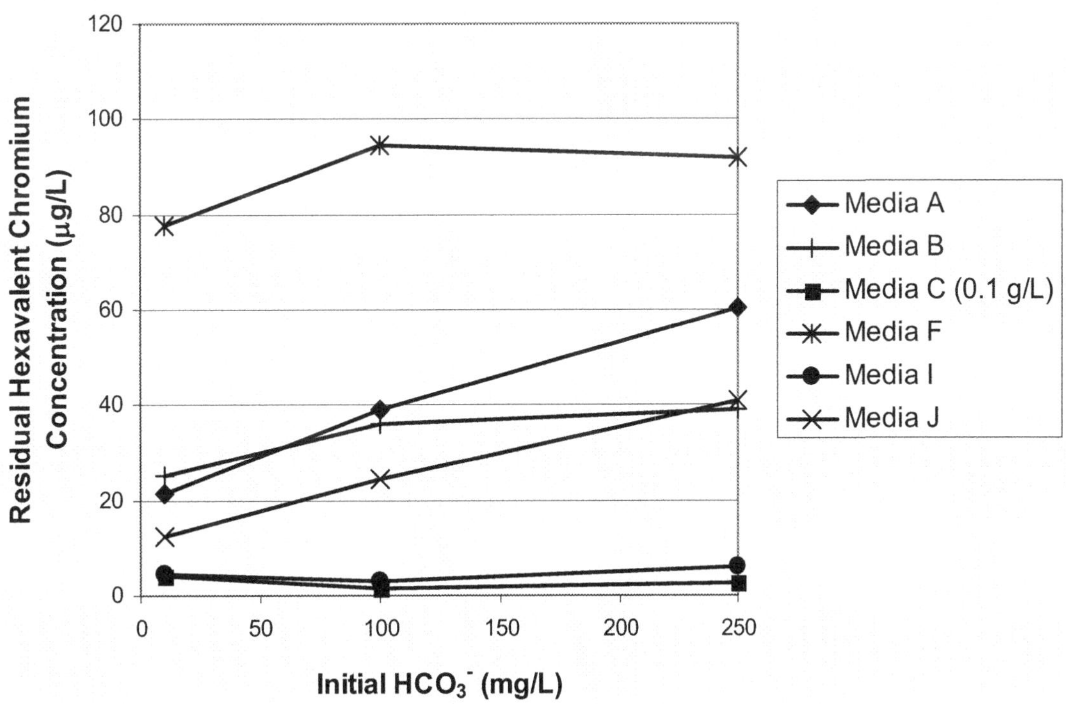

Figure 6.7 Effect of bicarbonate on hexavalent chromium removal

Table 6.3
Characterization of dual solute effects* on hexavalent chromium removal

Dual Solute	Media A	Media B	Media C	Media F	Media I	Media J
Chloride	- to =	NC†	=	-	- to =	- to =
Sulfate	-	NC	=	-	-	-
Bicarbonate	-	-	=	=	=	-
Silicate	- to =	=	=	- to =	=	=
TOC	-	-	=	-	=	-
Phosphate	=	NC	=	-	-	=
Nitrate	- to =	NC	=	- to =	=	=
Fluoride	- to =	NC	- to =	-	- to =	- to =
Arsenic	=	=	=	-	=	=
Perchlorate	=	=	=	=	=	=

* Dual solute effects on Cr(VI) removal:
- + indicates improved removal
- = indicates no effect
- – indicates diminished removal
- – to = indicates that the dual solute seems to slow the kinetics of Cr(VI) removal—Cr(VI) removal in the presence of the dual solute appears to be diminished at the 30-min sampling period, but diminished removal was not observed at the 2-hr sampling period

† NC indicates that the solute was not characterized.

Table 6.4
Average percent removal of dual solutes by treatment media

Media identification	Arsenic	Perchlorate	Nitrate
Media A	97%	-2%	-1%
Media C (0.1-g/L dose)	77%	4%	1%
Media F	57%	0%	1%
Media I	84%	93%	44%
Media J	45%	8%	3%

All of the treatment media evaluated in the dual solute experiments exhibited some ability to remove arsenic; however, media I was the only media that achieved significant removal of perchlorate and nitrate.

CONTINUOUS FLOW EXPERIMENTS

Ion Exchange

Conventional Anion Exchange Experiment

Four commercially available SBA exchange resins were evaluated for their ability to remove Cr(VI) from synthetic water in experiments using 0.5-in.-diameter columns. The objectives of these experiments were to: (1) compare breakthrough curves for Cr(VI) and sulfate, (2) measure effluent pH, and (3) compare the Cr(VI) removal performance of Type I and Type II SBA resins. The columns were operated at an EBCT of 2.5 min and a hydraulic loading rate of 1.0 gpm/sq ft, and they were run to exhaustion with respect to Cr(VI) removal.

Figure 6.8 plots breakthrough curves for the four resins. All of the resins were able to remove >95% of the influent Cr(VI) for at least 500–1,000 bed volumes. The resins' clear preference for Cr(VI) over the competing sulfate ion can be seen by comparing the breakthrough curves. With all four resins, influent sulfate equaled effluent sulfate after 500 bed volumes. This result indicates that each column's capacity for sulfate removal was exhausted long before its capacity for Cr(VI) removal and that the removal of Cr(VI) was preferred over sulfate removal. All resins exhibited a stable effluent pH after 500 bed volumes of operation. However, effluent pH did change over the first 500 bed volumes of operation. With the A300 resin, the initial effluent pH was elevated approximately 1.5 pH units, and with the A600 resin, it was depressed by about 1 pH unit.

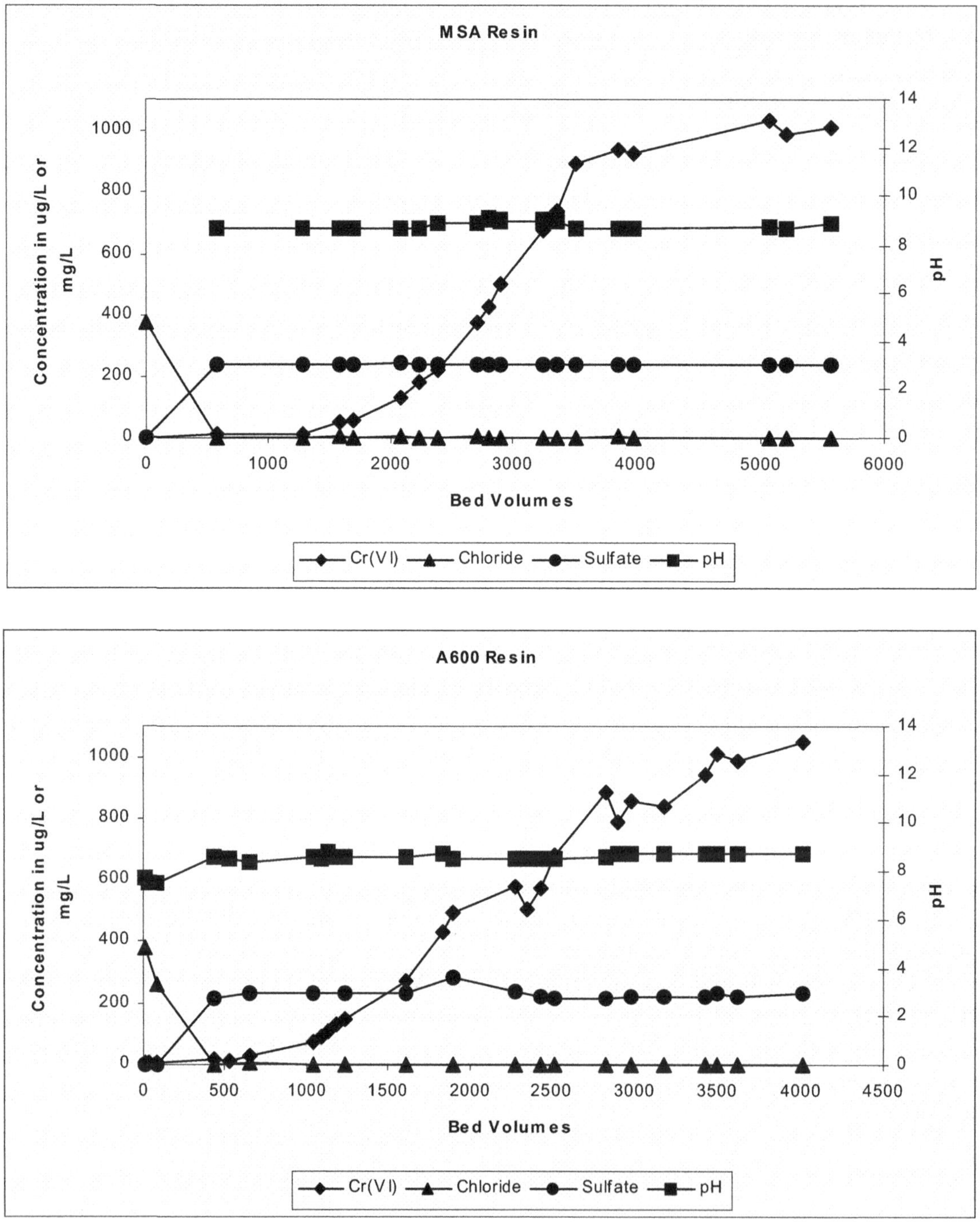

Figure 6.8 Breakthrough curves for strong base anion exchange resins

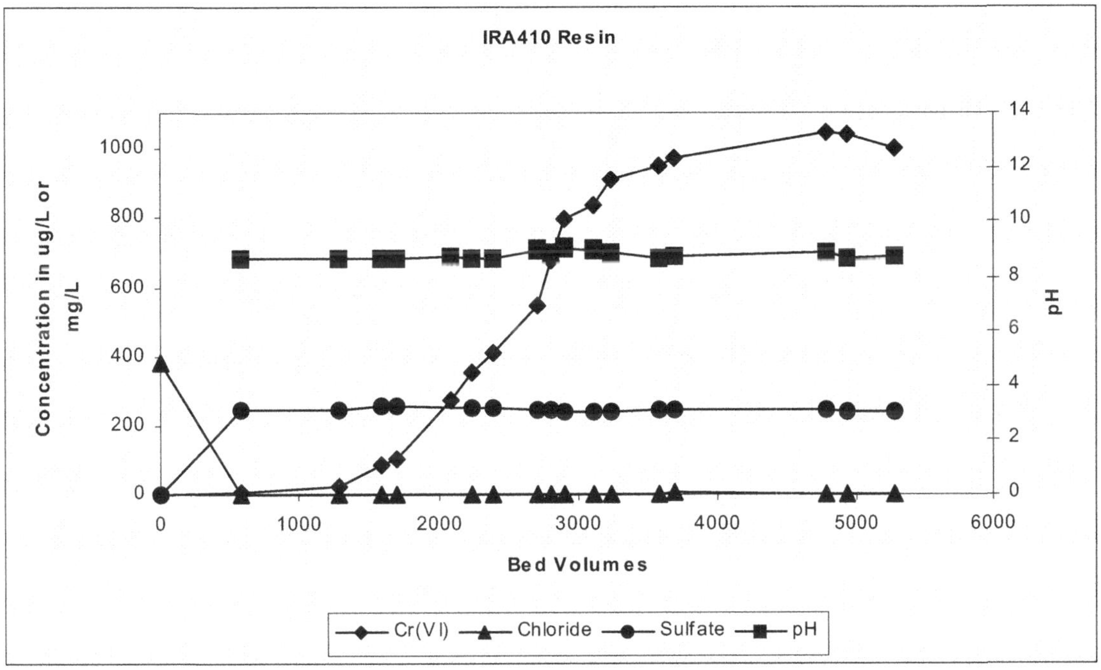

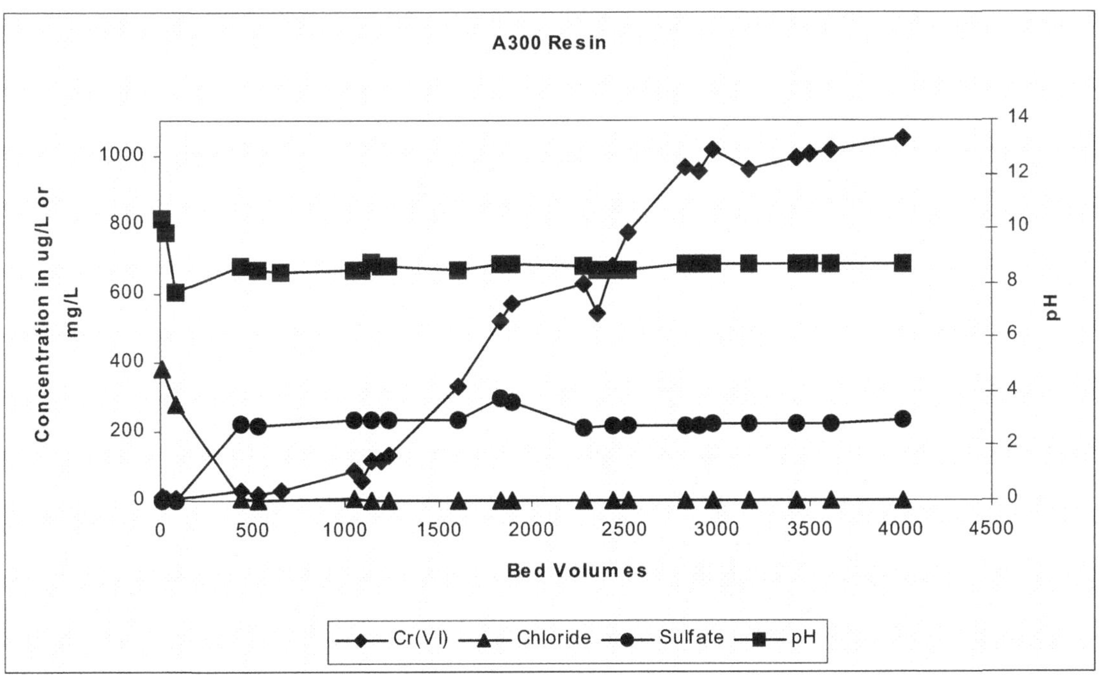

Figure 6.8 (Cont.)

The Cr(VI) breakthrough curves for the four resins are compared in Figure 6.9. The qualitative ranking of the resins' Cr(VI) removal capacity and the number of bed volumes to breakthrough were as follows:

A300 Type II, Gel	≈	A600 Type I, Gel	<	IRA410 Type II, Gel	<	MSA Type I, Macroporous

The Type I and Type II resins that were provided by the same manufacturer (A300 versus A600) performed nearly identically. Yet there was a difference in the performance of the Type I and Type II resins provided by various manufacturers (A600 versus MSA and A300 versus IRA410). Because of this wide variation in performance within and between classes of SBA resins, utilities that are considering an anion exchange process for Cr(VI) removal should evaluate a number of different resins. Of the resins tested, MSA was the best performer with respect to the number of bed volumes to breakthrough and exhaustion. Yet this resin had the lowest exchange capacity (1.1 meq/mL) of all those tested. The better performance of MSA in this limited set of tests indicates that a macroporous resin structure may be better for Cr(VI) removal than a gel structure.

In summary, these experiments indicated that SBA resins can be effective in controlling Cr(VI) to very low levels and that, for the resins tested, Cr(VI) removal is preferred over sulfate removal.

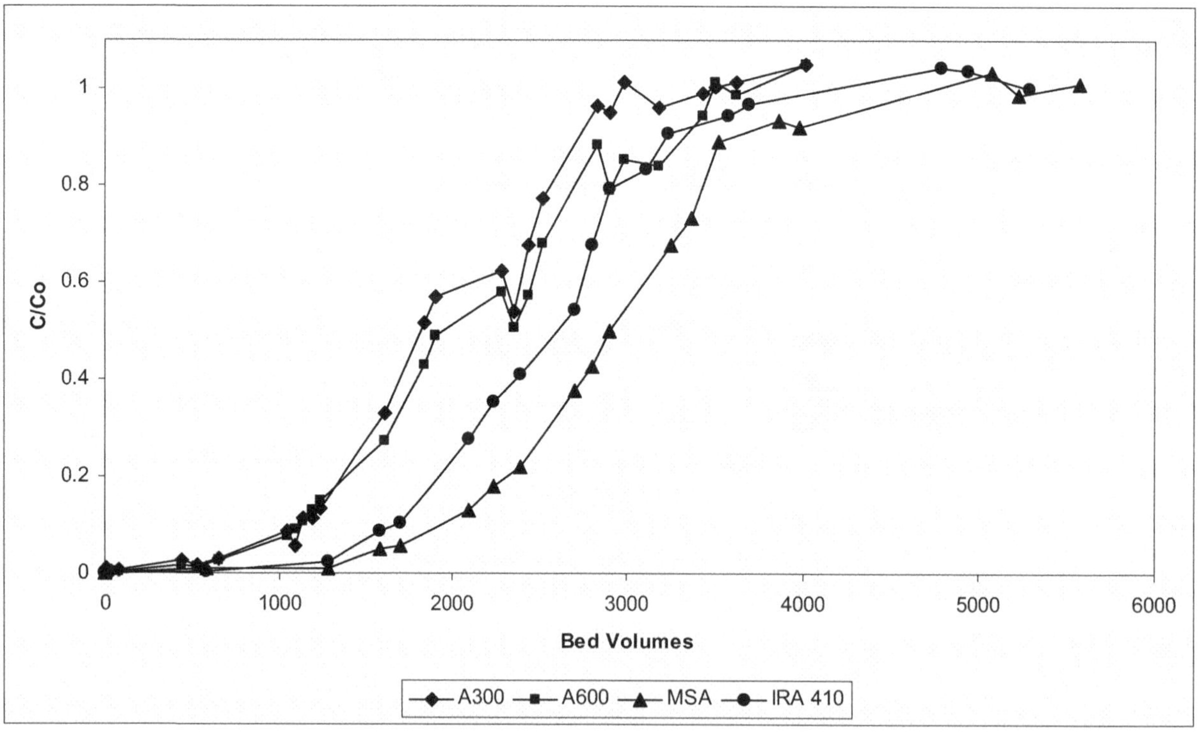

Figure 6.9 Comparison of resin performance

MIEX® Small-Scale Pilot Unit Experiment

A small-scale pilot unit was used to demonstrate Cr(VI) removal during continuous flow operation with the MIEX® treatment media. Two 5-day trials were conducted using the small-scale pilot unit. During both trials, the unit consistently removed >90% of the influent Cr(VI) concentration when operational conditions were maintained. Figure 6.10 shows the effluent Cr(VI) concentrations from the two 5-day trials conducted during 2003. As noted on the figure, the MIEX process did not reduce the Cr(VI) concentration below detection, nor did it consistently reduce the concentration below 5 μg/L. The shaded portions of Figure 6.10 indicate unit downtime; the dashed lines indicate regeneration events.

Sulfur-Modified Iron

The sulfur-modified iron treatment media displayed excellent Cr(VI) removal performance in all of the batch adsorption experiments. However, the batch adsorption experiments did not sufficiently characterize the media's performance under differing conditions. In order to accomplish this, mini-column experiments were conducted.

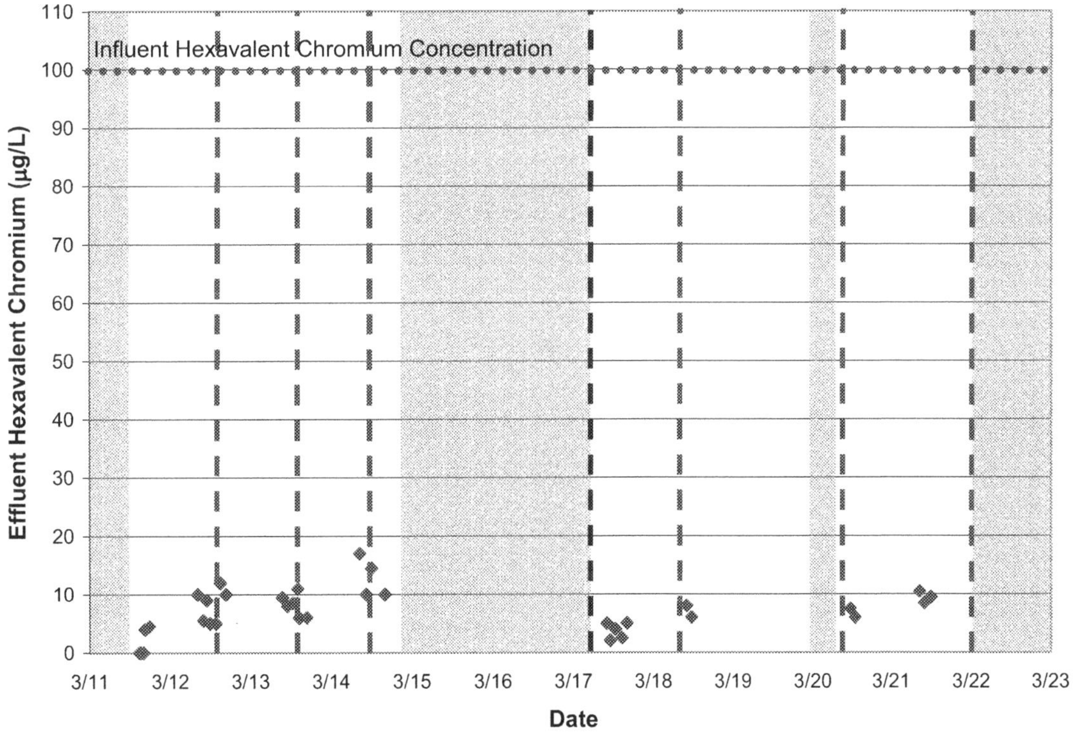

Figure 6.10 MIEX® Experimental results from small-scale pilot unit

Initial Capacity Column Experiment

A mini-column experiment was conducted to determine the Cr(VI) removal capacity of the sulfur-modified iron media in a short time frame. The ½-in.-diameter column was operated in an upflow configuration with a 2-min EBCT at a flow rate of 10 mL/min. The stock water supplied to the column was Milli-Q water buffered with 1 mM bicarbonate and containing 2.67 mg/L Cr(VI). The pH of the stock water was 8.3. Figure 6.11 shows the results of the initial capacity column experiment with the sulfur-modified iron treatment media.

The initial capacity column experiment indicated that the sulfur-modified iron treatment media had significant capacity to remove Cr(VI). Through 8,600 bed volumes of operation, the treatment media capacity was estimated to be 3.73 mg Cr(VI) per gram of media. Total chromium samples were taken throughout the experiment to ensure that the media was removing Cr(VI) and not just reducing it to Cr(III). As shown in Figure 6.11, results for the effluent total chromium sample match well with results for the Cr(VI) samples, indicating that the column was, in fact, removing chromium from the stock water.

A unique concern with the sulfur-modified treatment media is the release of iron during operation. During the initial capacity experiment, total iron samples were taken to track iron release. Iron release was highest during the initial 500 bed volumes of operation and then stabilized below 25 µg/L. Effluent pH was also measured during the initial column experiment. During this experiment, the column elevated the effluent pH to 1.5 pH units. The pH effect was evaluated further in subsequent experiments.

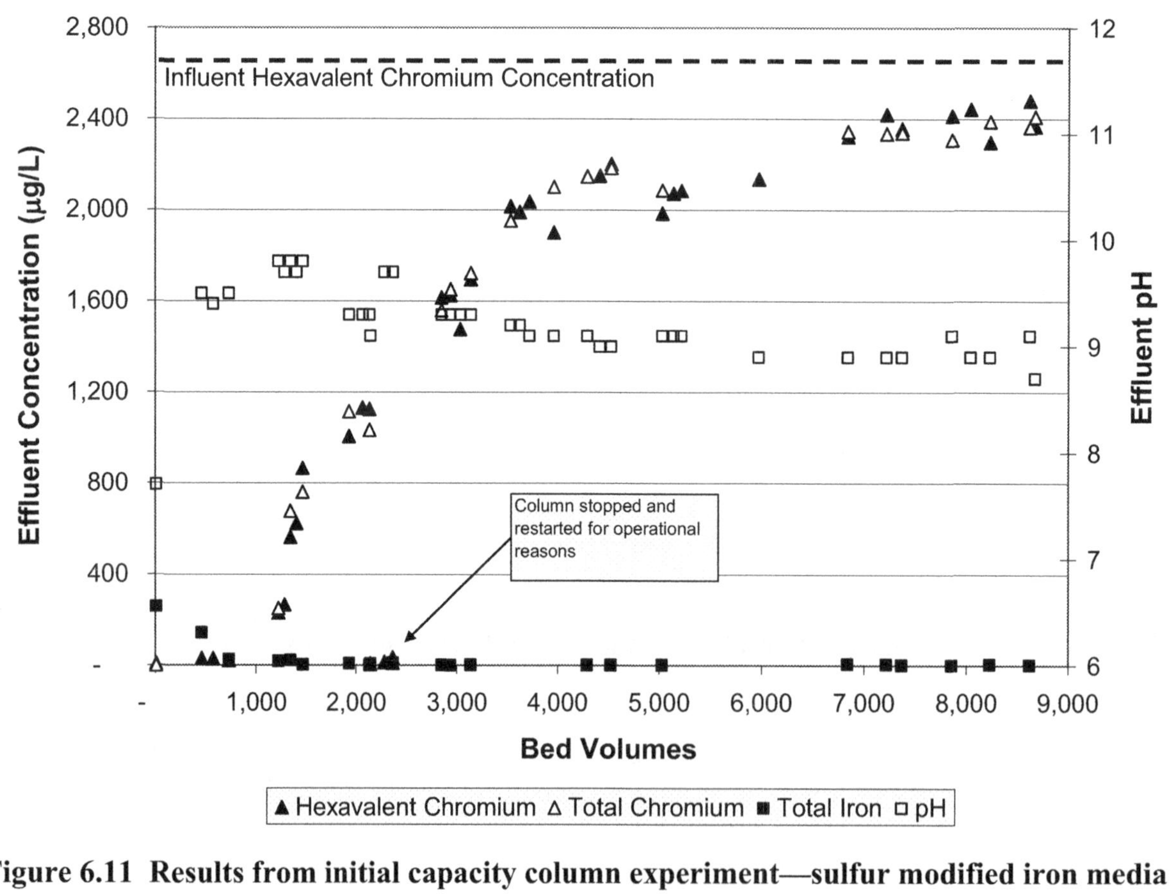

Figure 6.11 Results from initial capacity column experiment—sulfur modified iron media

pH Effect Column Experiment

The effect of pH on the column's Cr(VI) removal capacity was investigated after the initial capacity experiment. Two identical columns were operated in the same manner as the initial capacity experiment, with the exception of influent Cr(VI) concentration and pH. Both columns were operated with an influent Cr(VI) concentration of 1.00 mg/L. Column A was operated at an initial pH of 8.3; column B was operated at an initial pH of 7.0. Figure 6.12 shows the effluent Cr(VI) concentrations from columns A and B.

Both columns A and B were operated for more than 46,000 bed volumes. As shown in Figure 6.12, pH has a significant effect on Cr(VI) removal capacity. The breakthrough curve of column A was similar to that of the initial capacity column experiment, which was also run with an pH of 8.3. However, column B did not reach breakthrough in more than 46,000 bed volumes. Estimated Cr(VI) mass loadings on columns A and B after 46,000 bed volumes were 8.79 and 14.94 mg/g, respectively. These mass loading estimates were significantly higher than those observed in the initial capacity experiment.

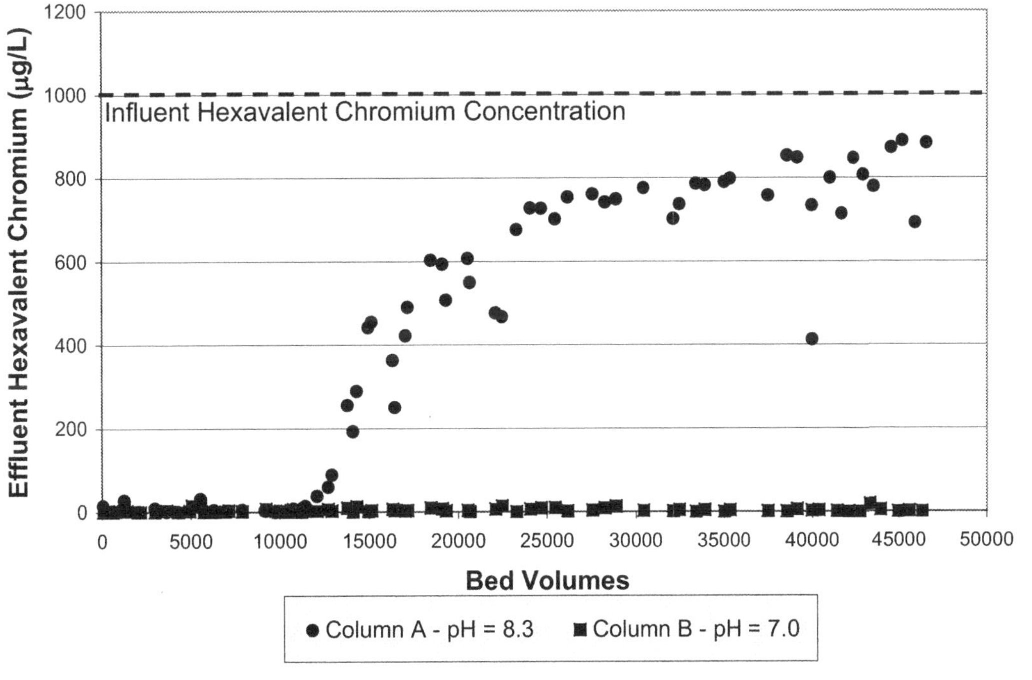

Figure 6.12 Results of column experiment on pH effect—sulfur-modified iron media

Confirmatory pH Effect Column Experiment

The confirmatory pH effect column experiment was conducted to confirm differences in the mass loading estimates of the initial capacity column experiment and the pH effect column experiment. The confirmatory pH effect column experiment was conducted in the same manner as the pH effect column experiment, except that the influent Cr(VI) concentration was increased to 2.67 mg/L (the same concentration used in the initial column capacity experiment). Column C, which served as a replicate of the initial capacity column experiment, was operated at an initial pH of 8.3. Column D was operated at an initial pH of 7.0. Figure 6.13 shows the results of the confirmatory pH effect column experiment.

Columns C and D were operated for more than 7,700 bed volumes. Column D, operated with an influent pH of 7.0, exhibited greater Cr(VI) removal capacity than column C, operated at an influent pH of 8.3. At 8,000 bed volumes, the estimated Cr(VI) mass loadings on columns C and D were 5.99 and 7.19 mg/g, respectively. Compared with the initial capacity column experiment, columns C and D exhibited much greater capacity for Cr(VI) removal. The reason for this difference in the removal capacity of the replicate columns is unknown.

Summary of Sulfur-Modified Iron Treatment Media Column Experiments

Five individual column experiments were conducted using the sulfur-modified treatment media. Operational conditions for each of the columns were the same. Two influent Cr(VI) concentrations and two influent pH conditions were evaluated. Table 6.5 summarizes the results of the column experiments.

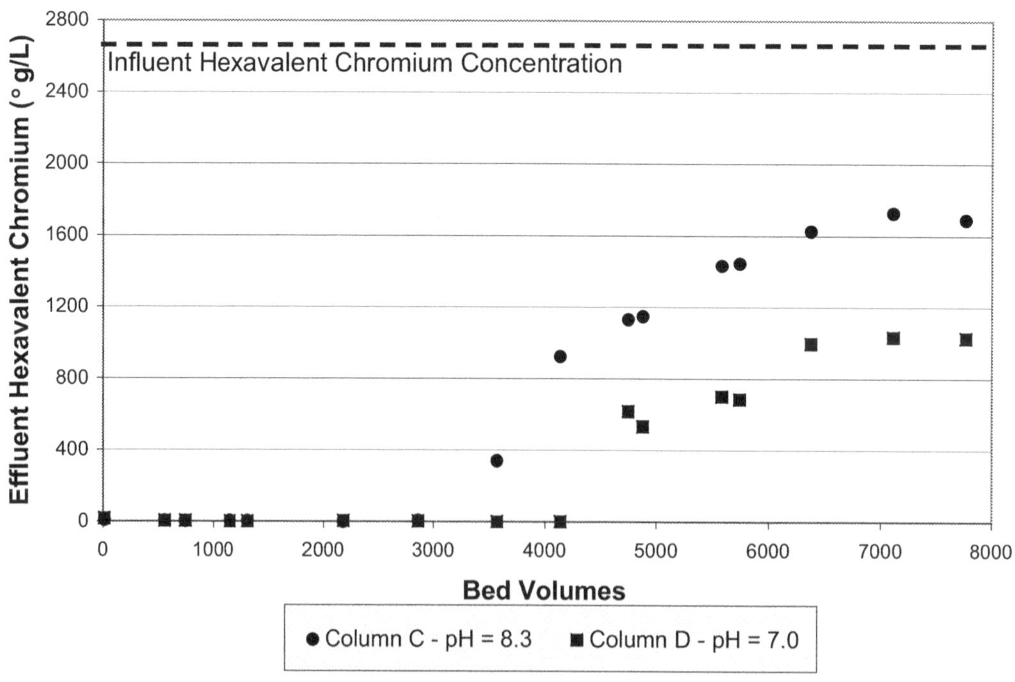

Figure 6.13 Results of confirmatory pH effect column experiment—sulfur-modified iron media

Table 6.5
Summary of column experiments on sulfur-modified iron treatment media

Column experiment	Influent Cr(VI) concentration (mg/L)	Influent pH	Bed-volume productivity by breakthrough (%)			Estimated mass loading (mg/g)
			10%	50%	90%	
Initial capacity	2.67	8.3	1,280	2,440	7,150	3.73
A	1.00	8.3	13,000	17,300	NE	8.79
B	1.00	7.0	NE	NE	NE	14.94
C	2.67	8.3	3,400	5,300	NE	5.99
D	2.67	7.0	4,400	NE	NE	7.19

NE = not exceeded

Results from the five sulfur-modified iron treatment media column experiments indicate two primary findings: lower influent Cr(VI) concentration and lower influent pH resulted in greater Cr(VI) removal capacity. Differences in bed-volume productivity between the initial capacity column and column D raise questions about the consistency of media performance. It is also not clear how the sulfur-modified treatment media would perform at influent Cr(VI) concentrations less than 100 μg/L, which would be more likely to occur in natural water supplies.

Batch isotherm experiments were conducted to help clarify the conflicting bed-volume productivity or capacity estimates developed in the sulfur-modified iron treatment media column experiments. Using the protocol developed for the batch adsorption experiments, these experiments replicated the experimental conditions of columns A and C to develop isotherms for the two influent Cr(VI) concentrations: 1.00 mg/L and 2.67 mg/L. Figure 6.14 shows the batch isotherm results in terms of media capacity and equilibrium Cr(VI) concentration.

Unfortunately, the batch isotherm experiments resulted in additional conflicting capacity estimates. Isotherms for the 1.00-mg/L and 2.67-mg/L Cr(VI) concentrations did not appear to exhibit consistent capacity across the range of equilibrium concentrations, indicating that adsorption may not have been the only removal mechanism taking place. The isotherm for the 2.67-mg/L Cr(VI) concentration indicates greater removal capacity than the isotherm for the 1.00-mg/L Cr(VI) concentration. This outcome contradicts the results of the column experiments.

Exhausted Media Leaching and Digestion Experiment—the TCLP and WET

Following the initial and confirmatory pH effect column experiments using sulfur-modified iron media, the columns of exhausted media were analyzed for their leaching potential and digested completely to determine chromium mass loading on the media. The columns were cut into quarter sections, labeled according to the upflow column operation: quarter section 1 at the bottom and quarter section 4 at the top. Media from each quarter section underwent total digestion, whereas column composite samples were used for leachate testing. Figure 6.15 shows results, by column, of the total digestion of each quarter section of media in terms of total chromium loading (micrograms of total chromium per gram of sulfur-modified iron media). The calculated total chromium loading, based on each column breakthrough curve, is also shown for comparison.

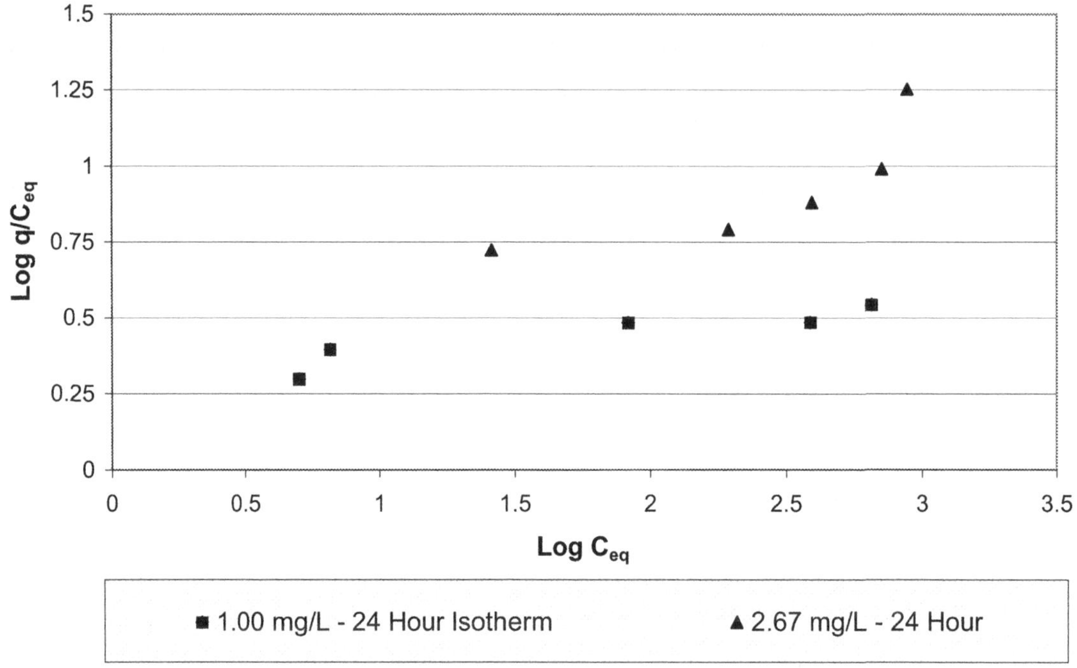

Figure 6.14 Results of batch isotherm tests with sulfur-modified iron treatment media

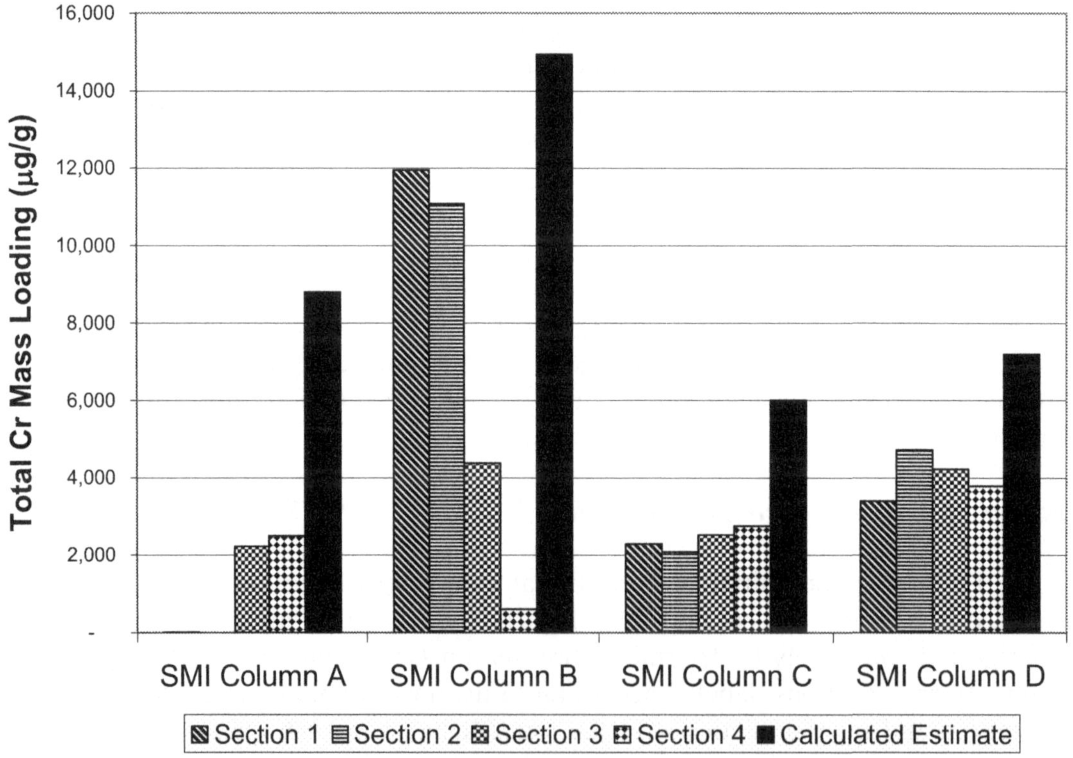

Figure 6.15 Results of total digestion of sulfur-modified iron treatment media

Mass loading results for the totally digested sulfur-modified iron media were clearly lower than the calculated estimates. It is unclear why not all of the chromium was recovered in the total digestion process. Perhaps some chromium or iron–chromium precipitate was not fully dissolved during the digestion process. In addition to the low recovery of chromium, column B was the only column to display increased chromium loading from the inlet to the outlet. Column B also exhibited the greatest chromium removal (>46,000 bed volumes without breakthrough), but in complete contrast, no chromium was found in the totally digested segments 1 and 2 of column A.

Table 6.6 gives results of the leachate testing on composite samples of each sulfur-modified iron column. The results are expressed as micrograms-per-liter of total chromium in the leachate.

Leachate testing results show that the TCLP does not cause the release of significant amounts of chromium. The WET, on the other hand, does cause significant amounts of chromium to be released from the media. Though these results are not conclusive, it appears that sulfur-modified iron would be unable to pass the WET evaluation.

Granular Ferric Hydroxide Column Experiment

The granular ferric hydroxide treatment media was evaluated in a column to confirm the results demonstrated in the bench-scale batch adsorption experiments. Influent water for the column was spiked with 100 μg/L Cr(VI) and arsenic. Figure 6.16 shows the effluent concentrations of both Cr(VI) and arsenic at the column midpoint and effluent locations. The EBCTs for the midpoint and effluent locations were 2.5 min and 5.0 min, respectively.

Similar to the results observed in the batch adsorption experiments, the granular ferric hydroxide treatment media exhibited a low capacity for removing Cr(VI). Cr(VI) breakthrough occurred at the effluent location at approximately 1,000 bed volumes. In contrast, arsenic breakthrough at the effluent location never occurred through 12,000 bed volumes of operation.

Table 6.6
Results of leachate testing in sulfur-modified iron columns

Column	Total chromium concentration of leachate (μg/L) TCLP	WET
A	1	31,320
B	31	88,633
C	65	22,247
D	0	40,280

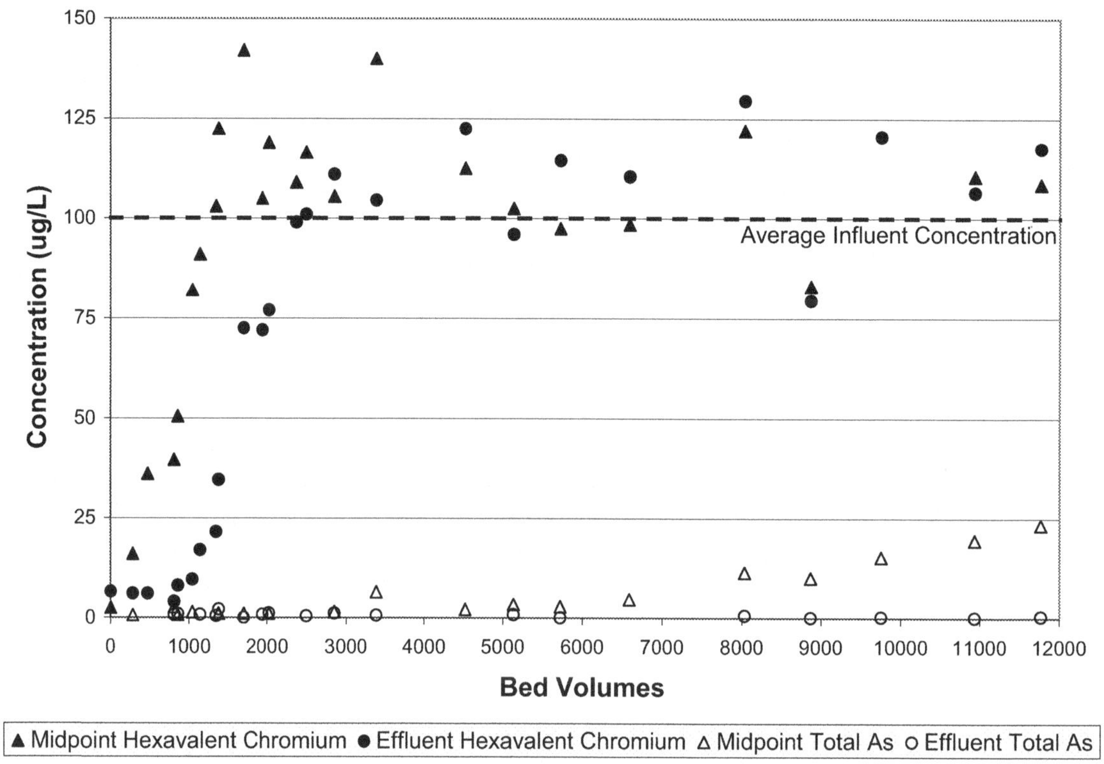

Figure 6.16 Results of column experiment with granular ferric hydroxide treatment media for removal of co-occurning solutes—Cr(VI) and arsenic

CHAPTER 7
MEMBRANE TECHNOLOGY: BENCH-SCALE RESULTS OF CHROMATE, ARSENATE, AND PERCHLORATE REJECTION BY NEGATIVELY CHARGED MEMBRANES

The primary focus of this chapter is the rejection of toxic anions by negatively charged high-pressure membranes. Although Cr(VI) is the focus of this report, the authors chose to evaluate chromate within the context of other toxic anions—arsenate and perchlorate—that have different solute properties (molecular weight, size, and charge). A transport model was used to provide a more mechanistic understanding of the rejection results. The chapter culminates with an assessment of Cr(VI) oxidative damage to polymeric membranes.

MEMBRANE PROPERTIES

As noted in Chapter 3, the membranes tested in this study included two RO membranes (LFC-1 and TFC-ULP), three NF membranes (TFC-S, ESNA, and MX07), and one tight UF membrane (GM).

Effect of pH and Conductivity on Zeta Potential

Zeta potential was measured to predict the effect of electrostatic exclusion on anion rejection resulting from charge repulsion between the anions (e.g., chromate, arsenate, and perchlorate) and the negatively charged membranes under varying pH conditions and in the presence of three electrolytes—potassium sulfate (K_2SO_4), potassium chloride (KCl), and calcium chloride ($CaCl_2$). Previous studies have shown that pH and mono- and divalent co- and counter-ions affect the surface charge of membranes (Childress and Elimelech 1996; Elimelech, Chen, and Waypa 1994). The (negative) surface charge of the membrane increases significantly with increasing pH. Membranes acquire a greater negative charge when pH increases and the same conductivity is maintained, regardless of the electrolyte added.

Zeta potential results obtained under varying pH and conductivity (ionic strength) conditions are shown in Figures 7.1 and 7.2. Under the same pH and conductivity conditions, the zeta potential of the membranes in the presence of the three electrolytes follows the order (negatively) $K_2SO_4 \geq KCl > CaCl_2$. However, the membranes exhibited less negative charge in the presence of electrolytes at relatively higher conductivity. Membrane zeta potential can be significantly affected by increasing conductivity, achieved through the addition of $CaCl_2$. The zeta potential of the membranes was reduced with increasing amounts of the divalent cation Ca^{2+}, as a result of ion bonding. However, in determining zeta potential with the co-ion Cl^-, bonding/adsorption can be reduced with increasing conductivity owing to a decrease in the electrical double layer. This mechanism can be applied to the divalent co-ion SO_4^{2-}. Bonding/adsorption of SO_4^{2-} can be reduced significantly even with increasing K_2SO_4 concentrations owing to a decrease in the electrical double layer because SO_4^{2-} bonding/adsorption occurs mostly as an inner-sphere complex. Moreover, the zeta potential of the membranes became less negative with increasing solution conductivity at the same pH, regardless of solution type. These results are consistent with the general observations of previous researchers (Childress and Elimelech 1996; Elimelech, Chen, and Waypa 1994).

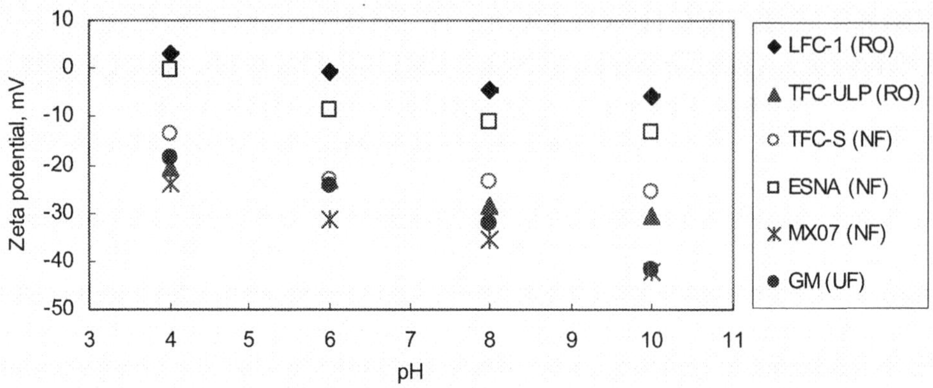

(A) pH 8 (with KCl)

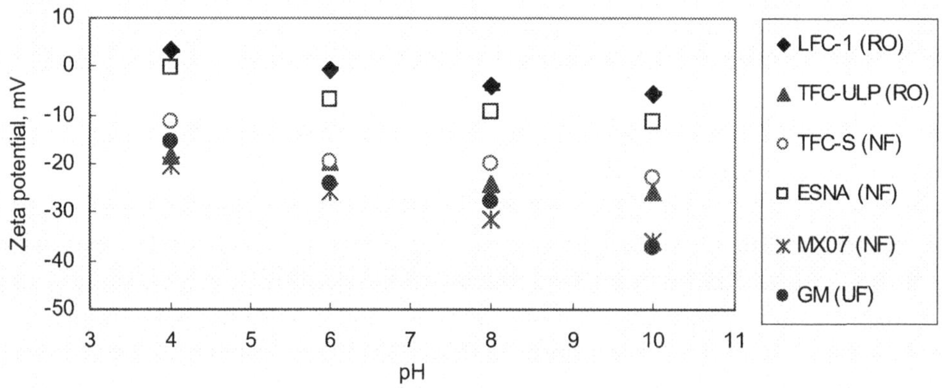

(B) pH 8 (with K_2SO_4)

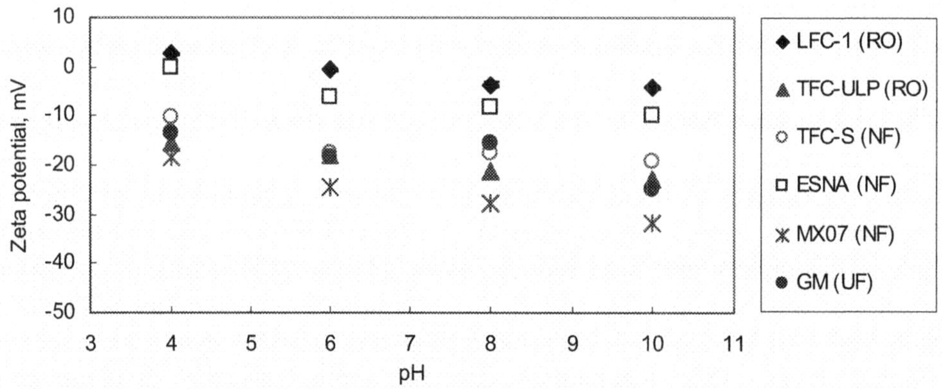

(C) pH 8 (with CaCl2)

Figure 7.1 Effect of pH on the zeta potential of RO, NF, and UF membranes

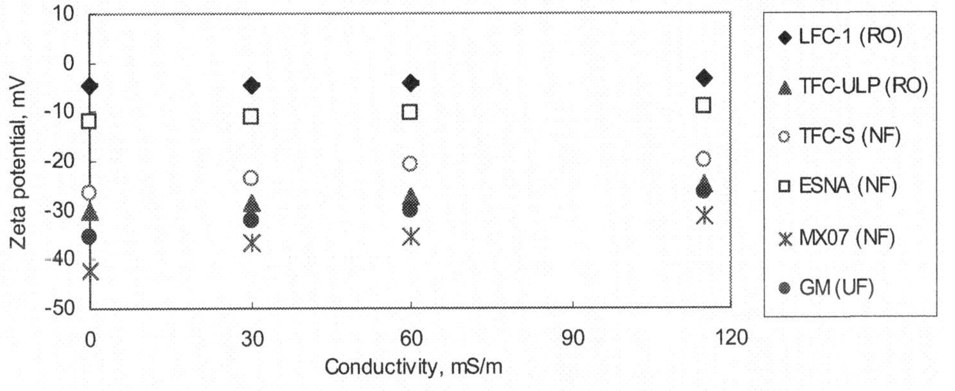

(A) Conductivity 30 mS/m (with KCl)

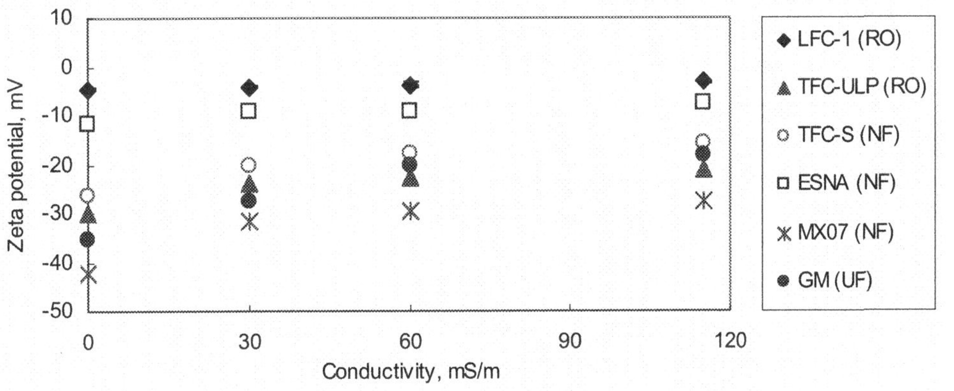

(B) Conductivity 30 mS/m (with K_2SO_4)

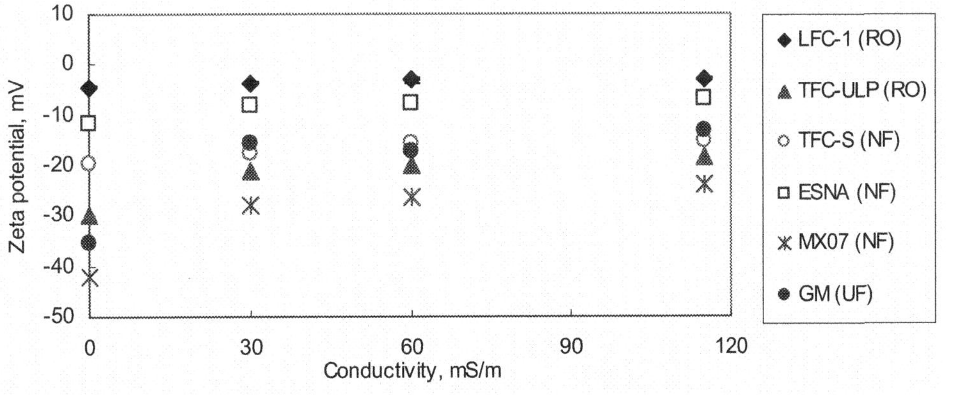

(C) Conductivity 30 mS/m (with $CaCl_2$)

Figure 7.2 Effect of conductivity on the zeta potential of RO, NF, and UF membranes

REJECTION OF CHROMATE, ARSENATE, AND PERCHLORATE UNDER CONDITIONS OF VARYING PH, CONDUCTIVITY, AND MEMBRANE ZETA POTENTIAL

One RO membrane (LFC-1), two NF membranes (ESNA and MX07), and one tight UF membrane (GM) were tested for their ability to reject three anions—chromate, arsenate, and perchlorate—in synthetic water under varying pH conditions (4, 6, 8, and 10) and conductivities (30, 60, and 115 mS/m). The equivalent conductivity of 100 mS/m is 1,000 microsiemens per centimeter, which is the unit typically used to describe conductivity in natural water supplies. If the typical conversion factor of 0.6 roughly applies, 100 mS/m is equivalent to about 600 mg/L TDS, which is in the general range of mineral concentrations in the natural water supplies tested in this study.Operating conditions are described in Chapter 3.

Figures 7.3 through 7.5 show rejection of the three anions in synthetic water at varying pH values and conductivities in the presence of the electrolytes K_2SO_4, KCl, and $CaCl_2$. Figures 7.3 and 7.4 show that rejection of the target anions by the ESNA, MX07, and GM membranes increased with increasing pH at constant conductivity (30 mS/m) and decreased with increasing conductivity at constant pH (pH 8); the RO (LFC-1) membrane achieved more than 90% rejection of the anions under all the experimental conditions. Figure 7.5 summarizes the rejection efficiencies observed in the presence of the three electrolytes and with a solution conductivity of 30 mS/m and a pH value of 8.

At constant pH and conductivity, anion rejection trends followed the electrolyte order $K_2SO_4 \geq KCl > CaCl_2$. These results can be explained by membrane zeta potential (see Figure 7.1). The surface charge of all the membranes tested became more negative with increasing pH at constant conductivity, and thus electrostatic repulsion of the anions became much stronger.

In terms of electrostatic interaction related to zeta potential (Figure 7.5), the Ca^{2+} binding causes the membrane surface charge to decrease significantly; thus, the target anions are rejected less in the presence of $CaCl_2$ than in the presence of KCl or K_2SO_4. Rejections of the three anions in the presence of K_2SO_4 are slightly higher than in the presence of KCl. These results can be supported on the basis of electrostatic interaction with measured zeta potential results (Figures 7.1 and 7.2). It was assumed that zeta potentials with K_2SO_4 would have higher negative values than those with KCl. However, zeta potential measurements showed that membrane surface charge in the presence of K_2SO_4 had slightly more negative values than membrane surface charge in the presence of KCl. These results can be explained by increasing polarization of the KCl and K_2SO_4 concentrations at the interface of the membranes (Yoon 2001, Elimelech and O'Melia 1990). According to these researchers, K_2SO_4 concentration increased more than KCl concentration over time at the membrane interface during filtration. These researchers also observed that zeta potential at the membrane interface, where solute concentration increases, can be different from what is obtained with the bulk concentration. Because more polarization occurs with increasing K_2SO_4 concentrations, greater reduction of the electrical double layer can be achieved by increasing K_2SO_4 than by increasing KCl. Thus, zeta potential at the membrane interface during filtration may show similar negative values in the presence of both electrolytes.

As shown in Figure 7.4, rejection of the three target anions decreased with increasing solution conductivity. These results can be explained by a reduction in electrostatic repulsion and an increase in solute (anion) transport through negatively charged membranes with increasing conductivity. In terms of electrostatic repulsion, the electrical double layer is reduced with increasing conductivity, and thus the zeta potential becomes less negative at the inner sphere of the membrane (Figure 7.2). This reduction in the electrical double layer may favor partitioning and diffusion, which affect solute transport through the negatively charged membrane.

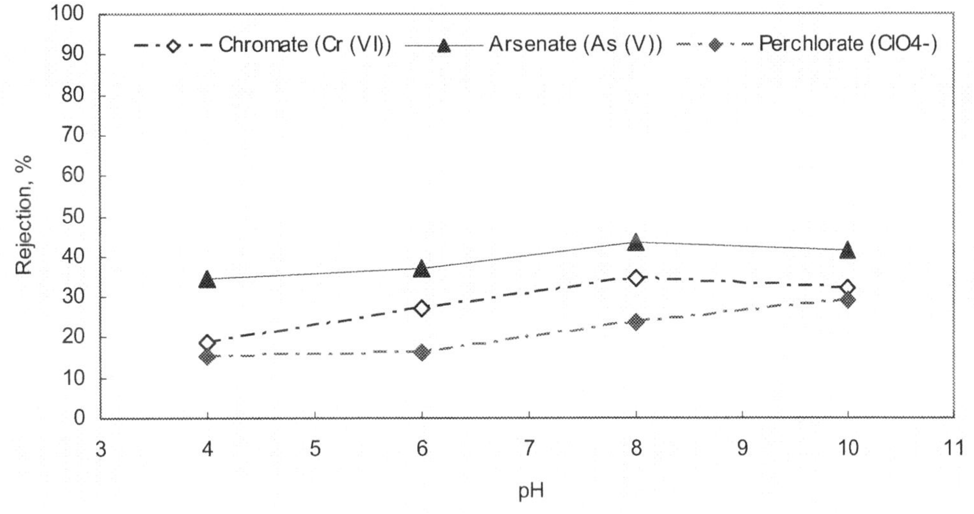

(A) GM (UF membrane)

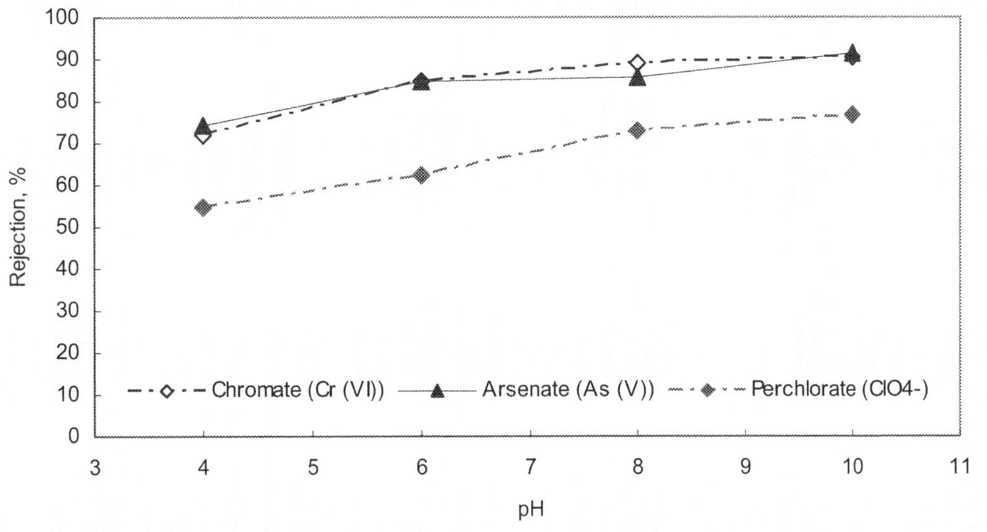

(B) ESNA (NF membrane)

Figure 7.3 Effect of pH on membrane rejection of chromate, arsenate, and perchlorate at a solution conductivity of 30 mS/m (conductivity adjusted with KCl)

(continued)

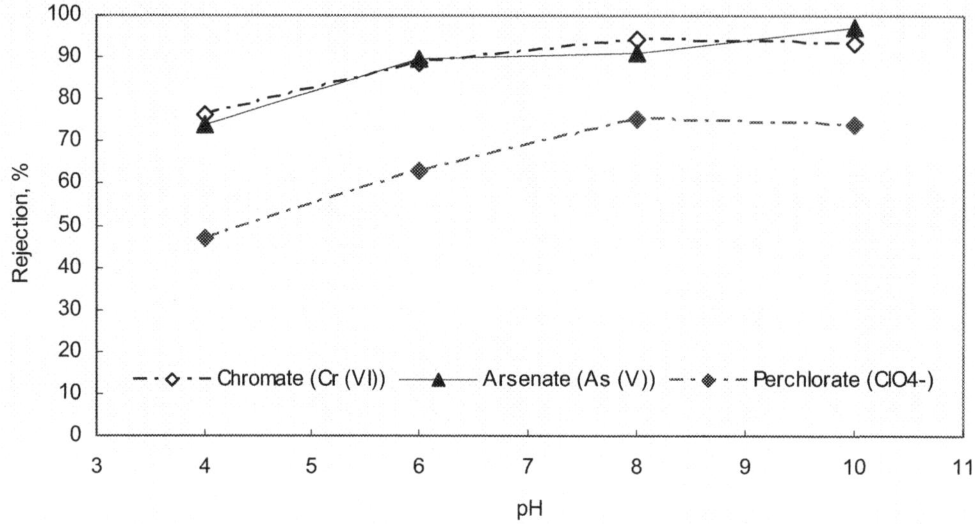

(C) MX07 (NF membrane)

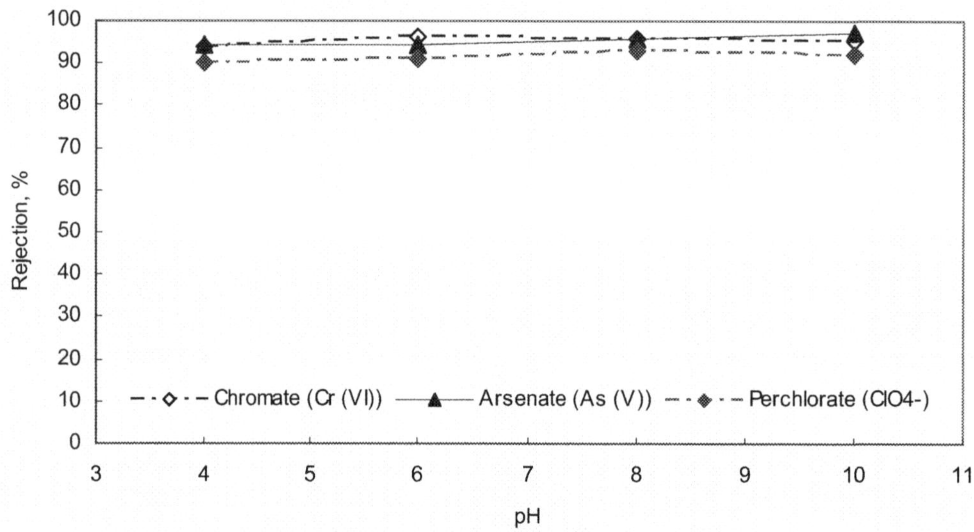

(D) LFC-1 (RO membrane)

Figure 7.3 (Continued)

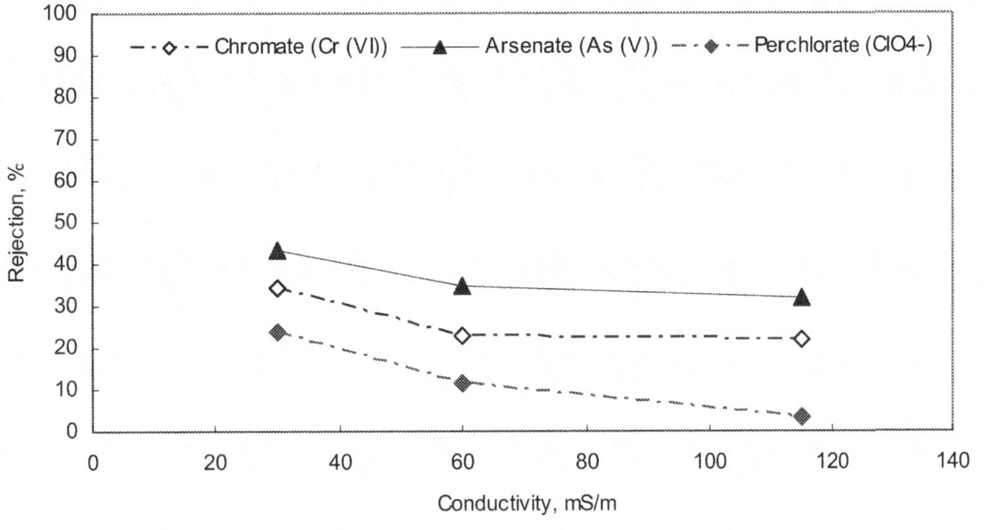

(A) GM (UF membrane)

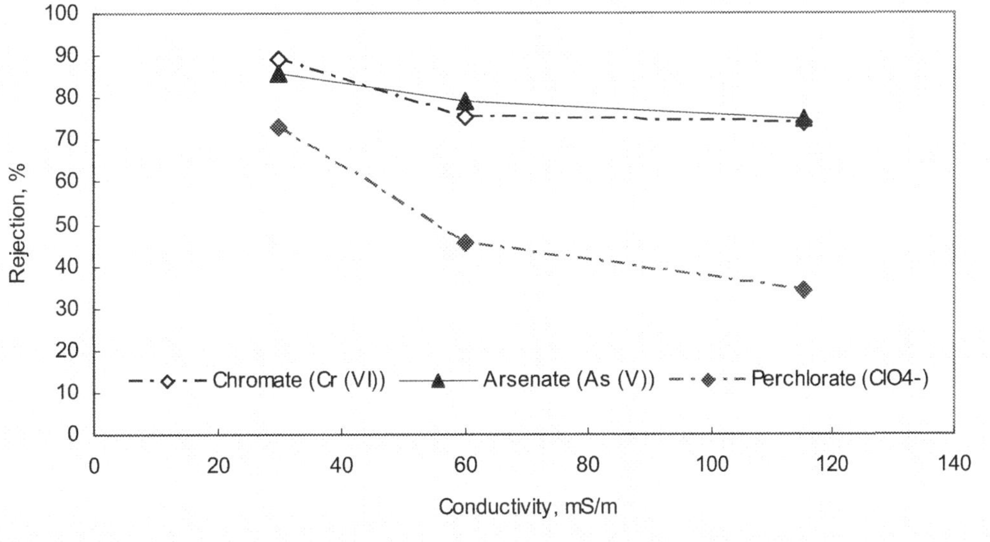

(B) ESNA (NF membrane)

Figure 7.4 Effect of conductivity on membrane rejection of chromate, arsenate, and perchlorate at pH 8 (conductivity adjusted with KCl)

(continued)

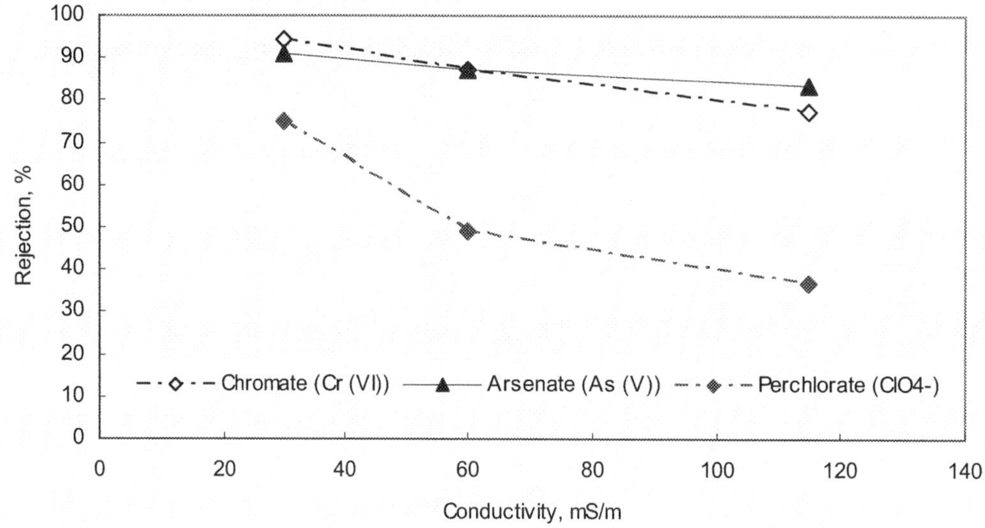

(C) MX07 (NF membrane)

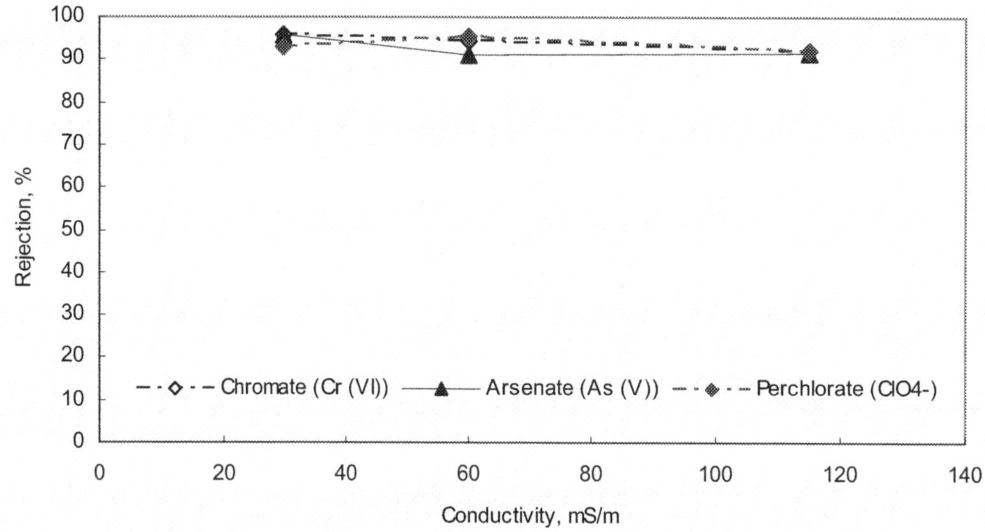

(D) LFC-1 (RO membrane)

Figure 7.4 (Continued)

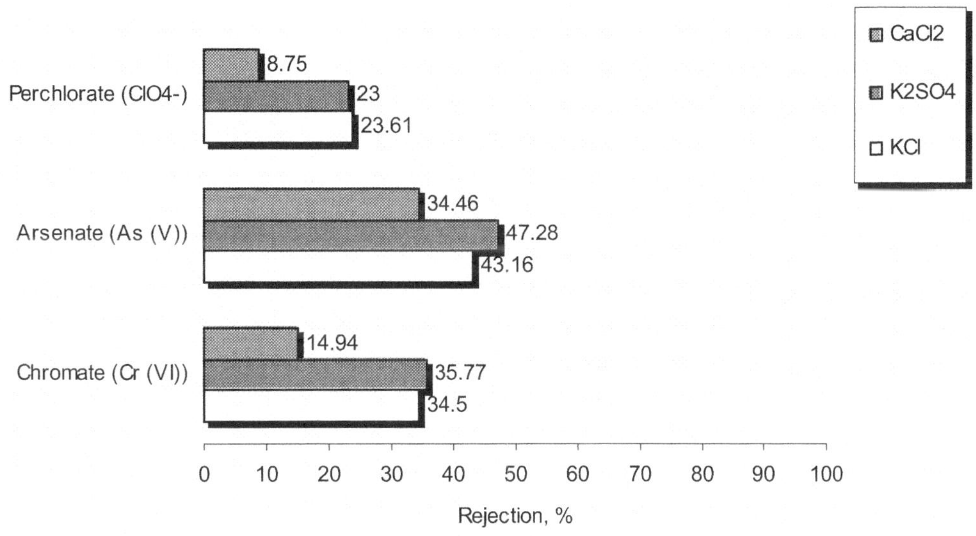

(A) GM (UF membrane)

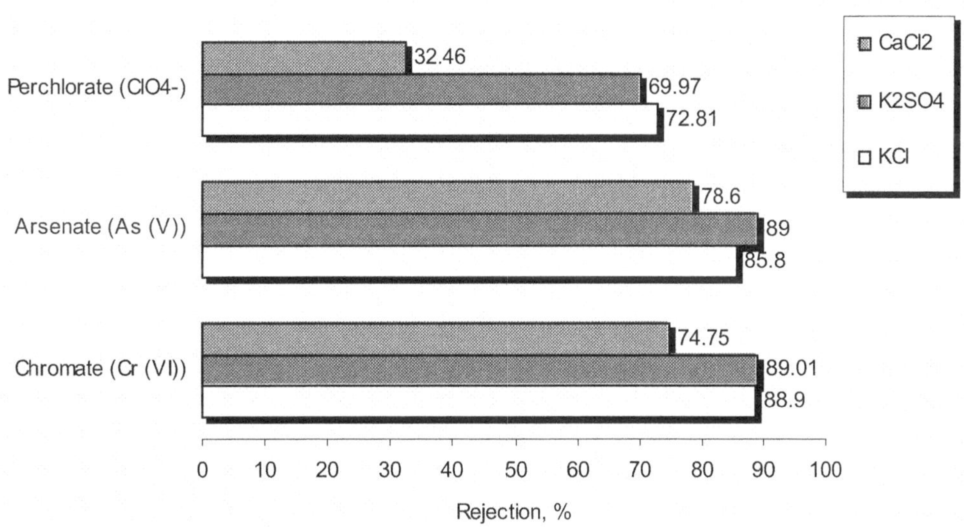

(B) ESNA (NF membrane)

Figure 7.5 Effect of electrolytes on membrane rejection of chromate, arsenate, and perchlorate at pH 8 and a solution conductivity of 30 mS/m

(continued)

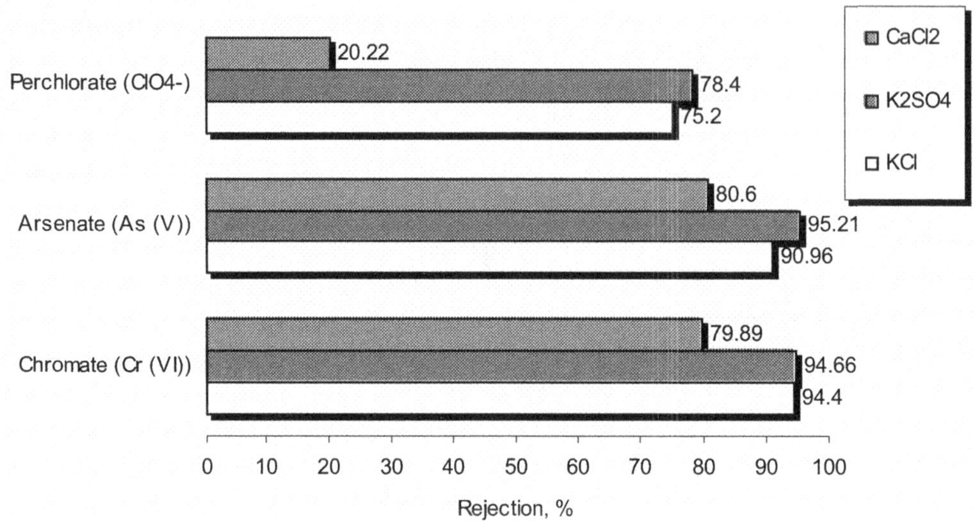

(C) MX07 (NF membrane)

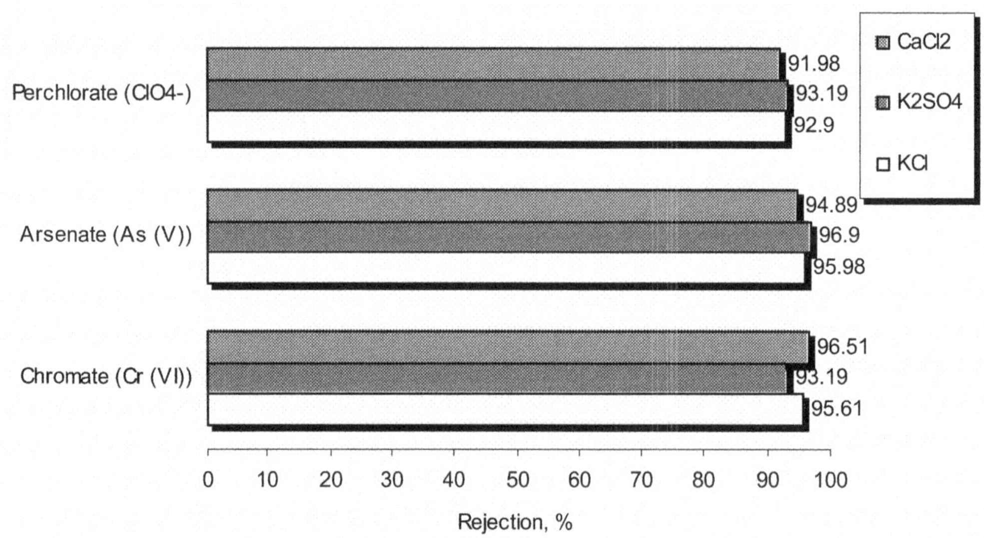

(D) LFC-1 (RO membrane)

Figure 7.5 (Continued)

Effect of Co- and Counter-Ion Rejection on Anion Transport in Synthetic Water

During the membrane filtration runs, each co- and counter-ion—SO_4^{2-}, Cl^-, K^+, and Ca^{2+}—in the feed and permeate was measured to determine how the anions at the interface of the membranes might change. Figures 7.6 and 7.7 show the co- and counter-ion rejection capabilities of four membranes (LFC-1, ESNA, MX07, and GM). The measurements were conducted using three electrolytes—K_2SO_4, KCl, and $CaCl_2$—at varying pH and conductivity under the same experimental conditions, including recovery and J_o/k ratio. As depicted in Figures 7.6 and 7.7, the efficiency of each membrane's rejection of co- and counter-ions varied from 0% to 95%, depending on pH and conductivity. The increase in ion concentration because of the reduction of the electrical double layer can affect charge density at the membrane interface and significantly influence anion transport through the membrane. As shown in both figures, the NF and UF membranes were more effective at rejecting the divalent co-ion SO_4^{2-} than the monovalent co-ion Cl^- in the presence of K^+ at varying pH and conductivity. Also, rejection of the counter-ion K^+ was greater in the presence of SO_4^{2-} than in the presence of Cl^-. However, all rejections were higher (~90%) under all the experimental conditions, regardless of the effect of co- and counter-ions.

Effect of pH and Conductivity on As(III) Rejection and Membrane Zeta Potential

As(III) exists as the uncharged species H_3AsO_4 in natural water supplies (pK_{a1}=9.13), whereas As(V) exists as the anionic forms $H_2AsO_4^-$ and $HAsO_4^{2-}$ in natural water supplies (pK_{a1}=2.19, pK_{a2}=6.94) (Sato et al. 2002). For this reason, As(III) was used as a surrogate solute to verify the rejection mechanisms of electrostatic exclusion in NF and UF membranes and size exclusion in the RO membrane.

These experiments were performed under the same pH, and conductivity (adjusted with KCl) conditions as for the three anions tested. Figure 7.8 shows As(III) rejection at varying pH and conductivity values. As(III) rejection by the RO membrane (LFC-1) was greater than 90% under all conditions because size exclusion is the dominant rejection mechanism in RO membranes. As(III) rejection by the NF and UF membranes did not change significantly at pH 4, 6, or 8 or at a solution conductivity of 30, 60, or 115 mS/m because As(III) exists predominantly as an uncharged species below pH 9.13. However, rejections by NF (ESNA and MX07) and UF (GM) membranes increase at pH 10 when As(III) exists as a charged species. At pH 10, the MX07 membrane shows slightly higher As(III) rejection than the ESNA membrane, even though ESNA provides higher rejection than MX07 at pH 4–8. These results support the premise that membrane surface charge is an important mechanism determining anion rejection by negatively charged membranes. Based on the results of zeta potential measurements under various conditions (Figures 7.1 and 7.2), MX07 shows more negative values than ESNA.

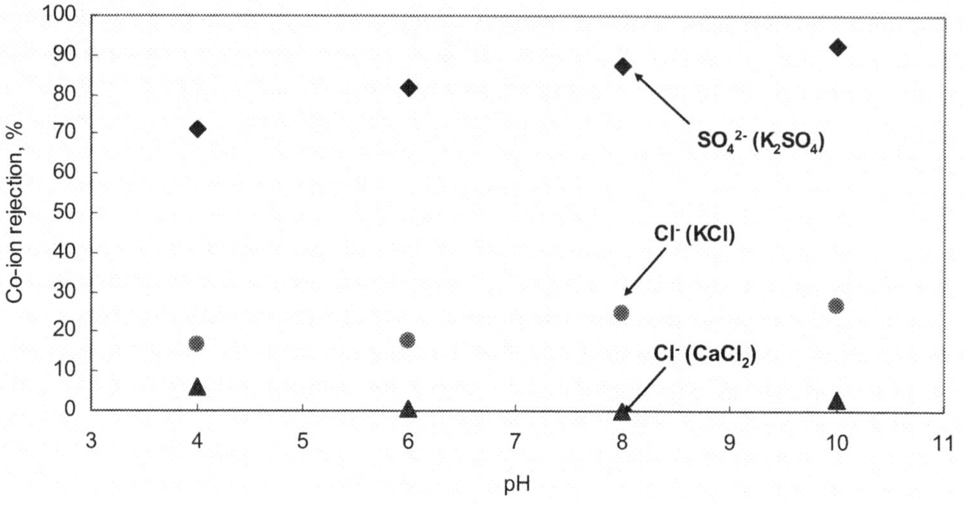

(A) Co-ion rejection by GM (UF membrane)

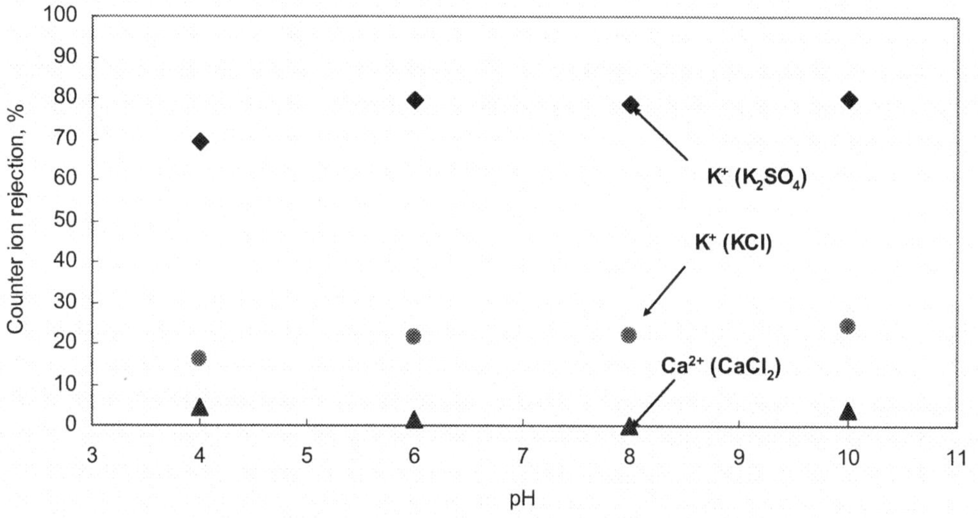

(B) Counter-ion rejection by GM (UF membrane)

Figure 7.6 Effect of pH on membrane rejection of co- and counter-ions at a solution conductivity of 30 mS/m

(continued)

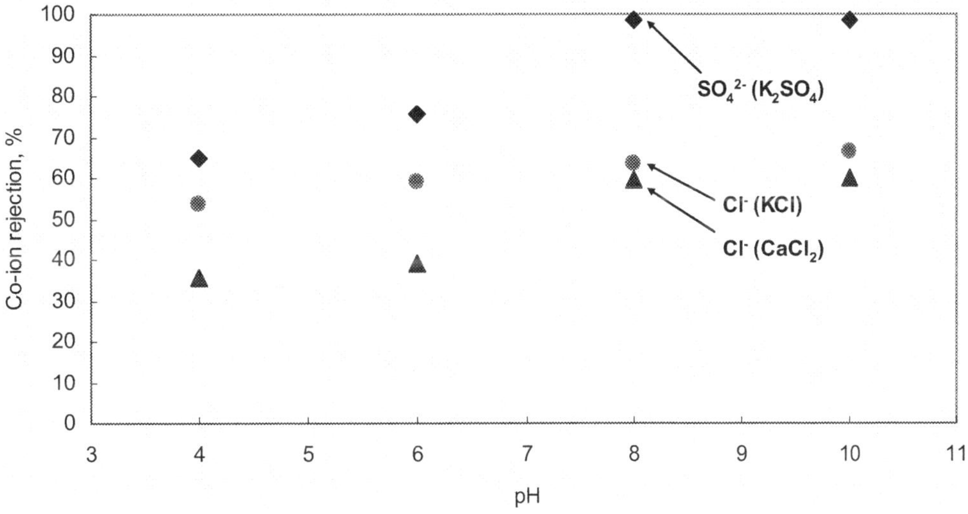

(C) Co-ion rejection by ESNA (NF membrane)

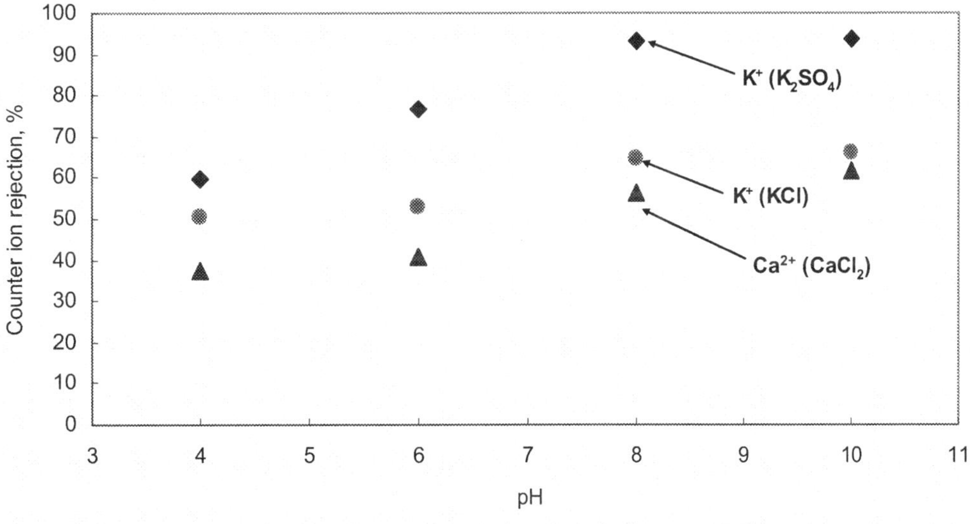

(D) Counter-ion rejection by ESNA (NF membrane)

Figure 7.6 (Continued)

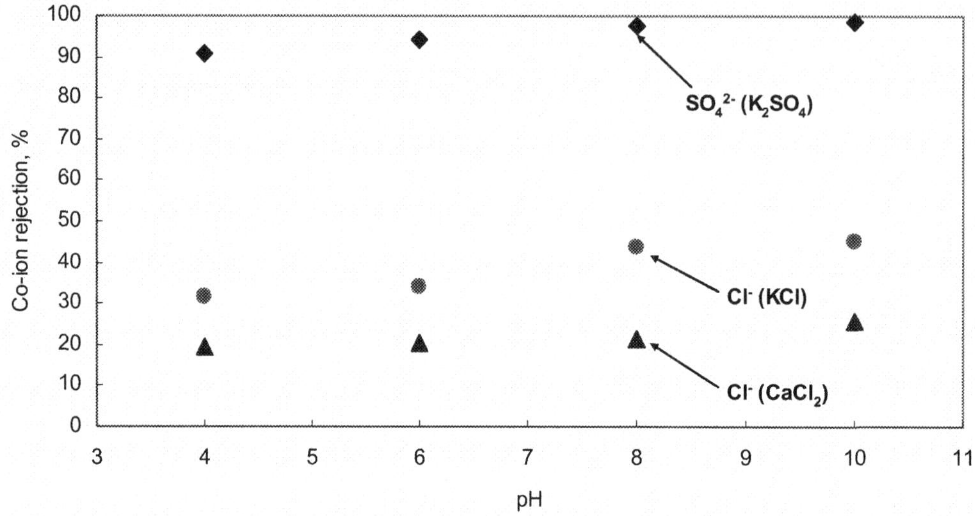

(E) Co-ion rejection by MX07 (NF membrane)

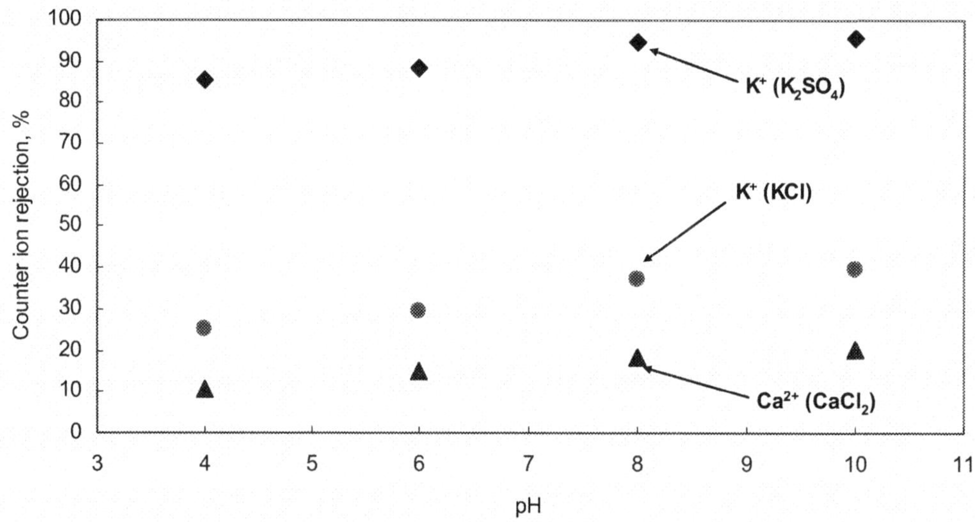

(F) Counter-ion rejection by MX07 (NF membrane)

Figure 7.6 (Continued)

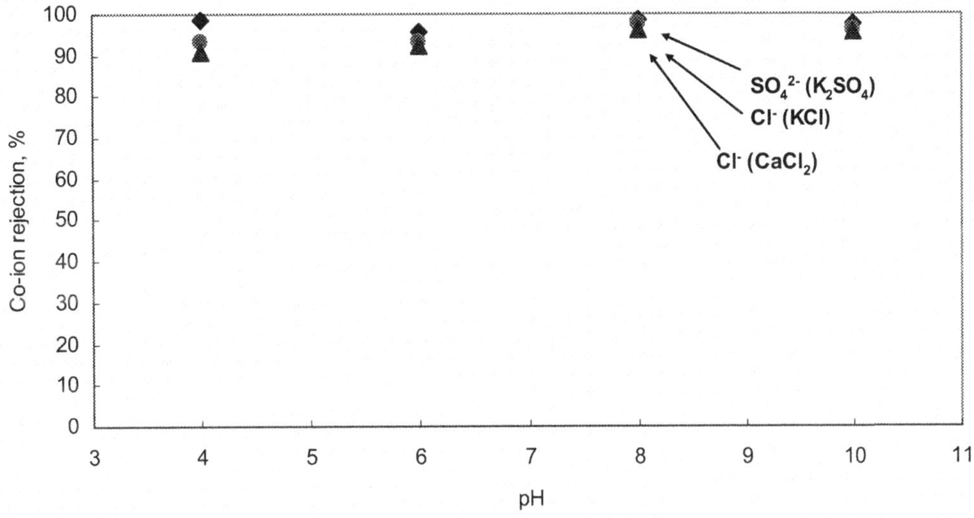

(G) Co-ion rejection by LFC-1 (RO membrane)

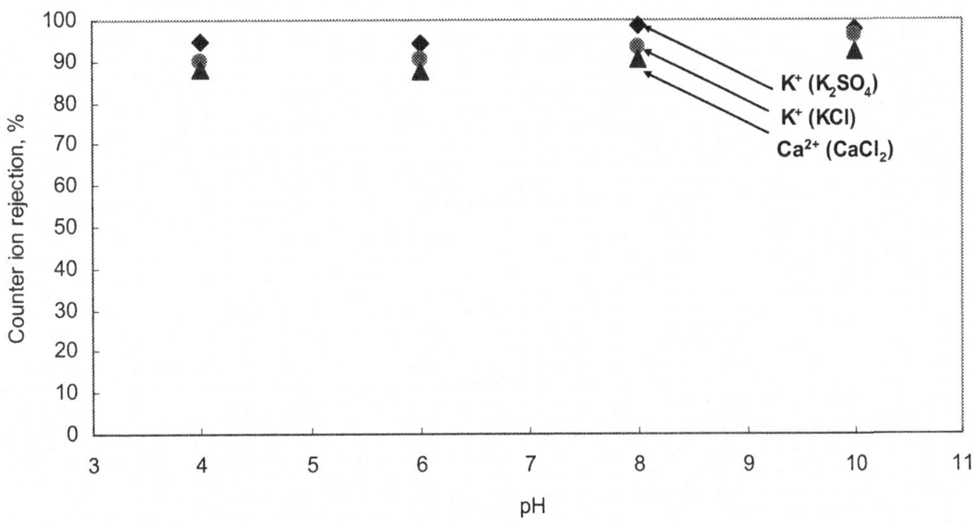

(H) Counter-ion rejection by LFC-1 (RO membrane)

Figure 7.6 (Continued)

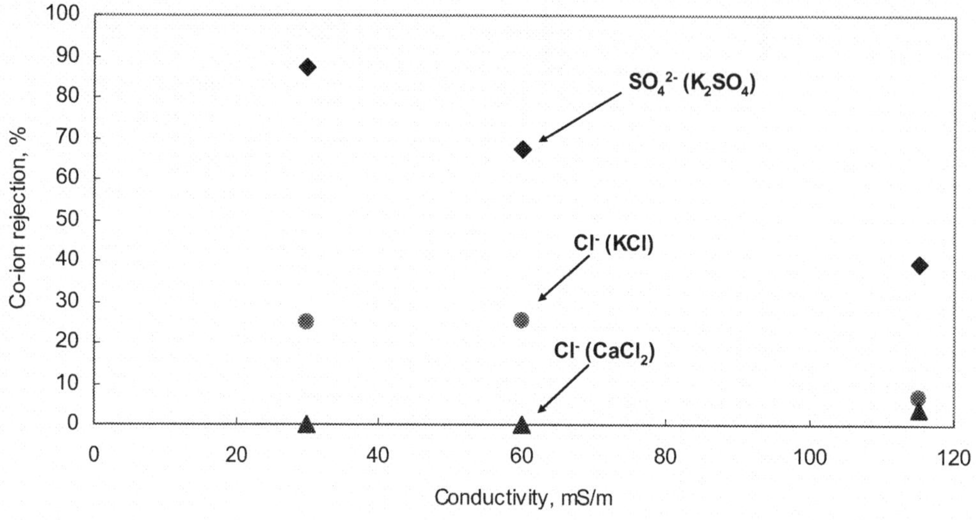

(A) Co-ion rejection by GM (UF membrane)

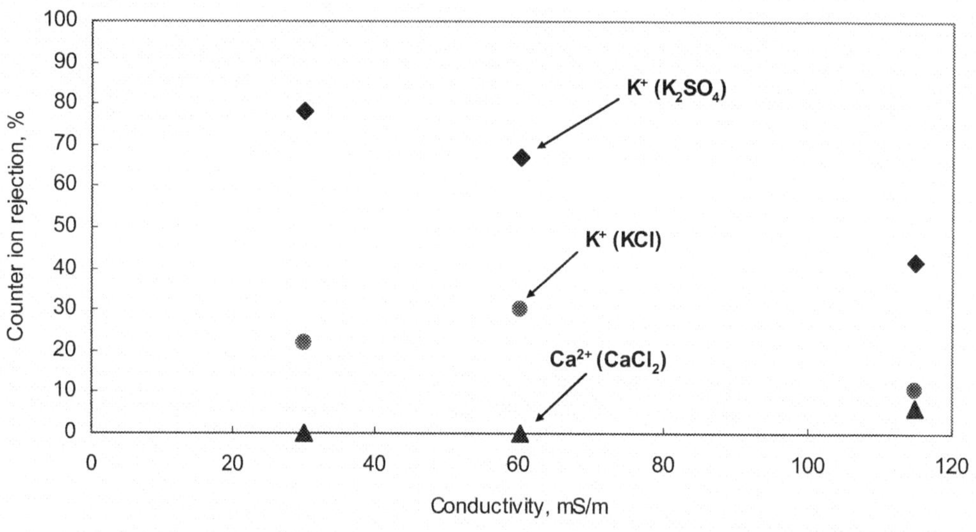

(B) Counter-ion rejection by GM (UF membrane)

Figure 7.7 Effect of conductivity on membrane rejection of co- and counter-ions at pH 8

(continued)

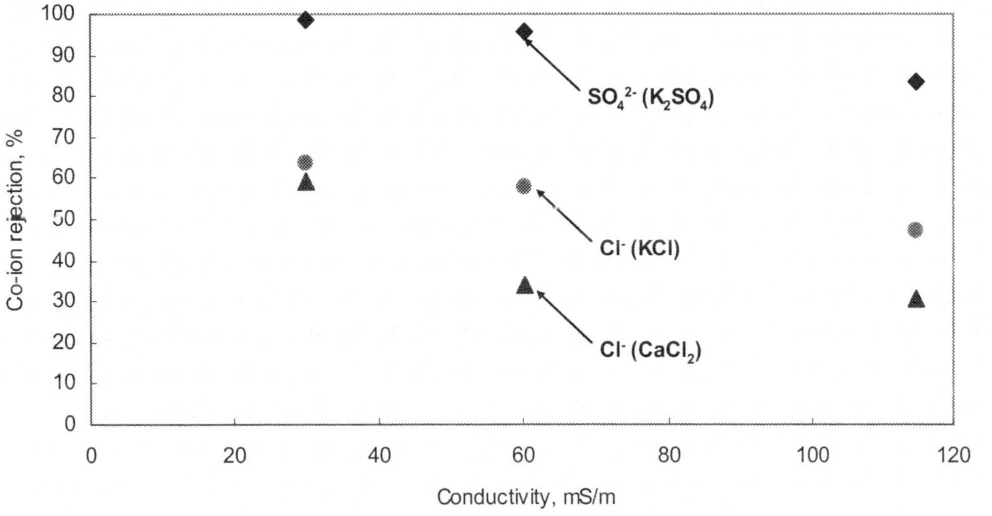

(C) Co-ion rejection by ESNA (NF membrane)

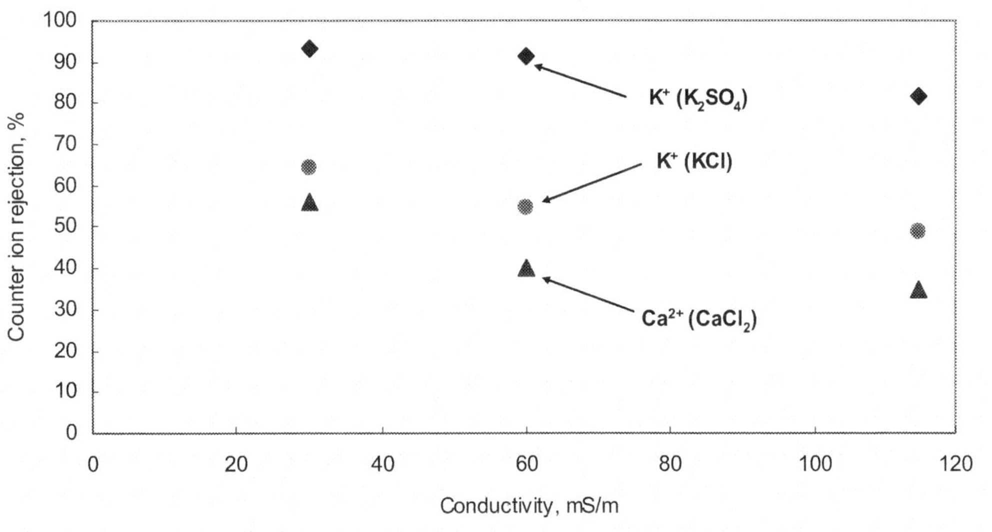

(D) Counter-ion rejection by ESNA (NF membrane)

Figure 7.7 (Continued)

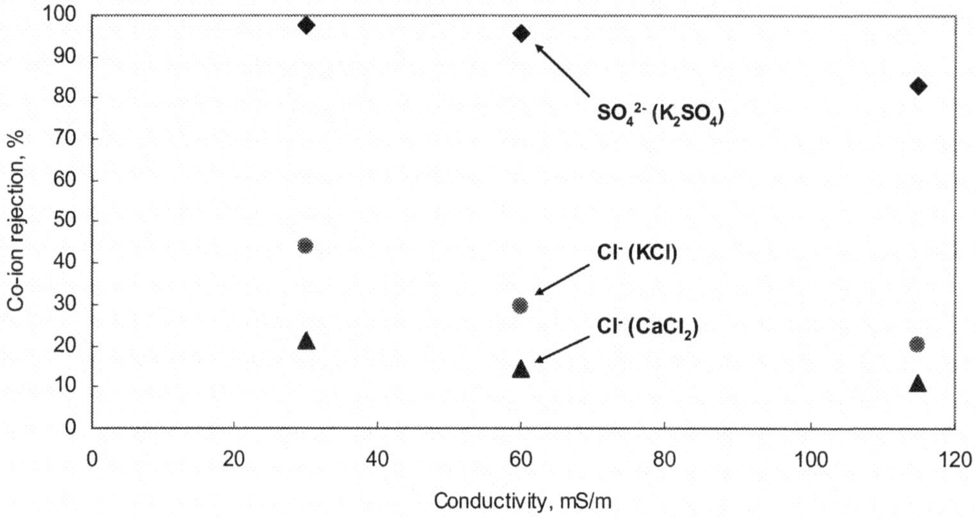

(E) Co-ion rejection by MX07 (NF membrane)

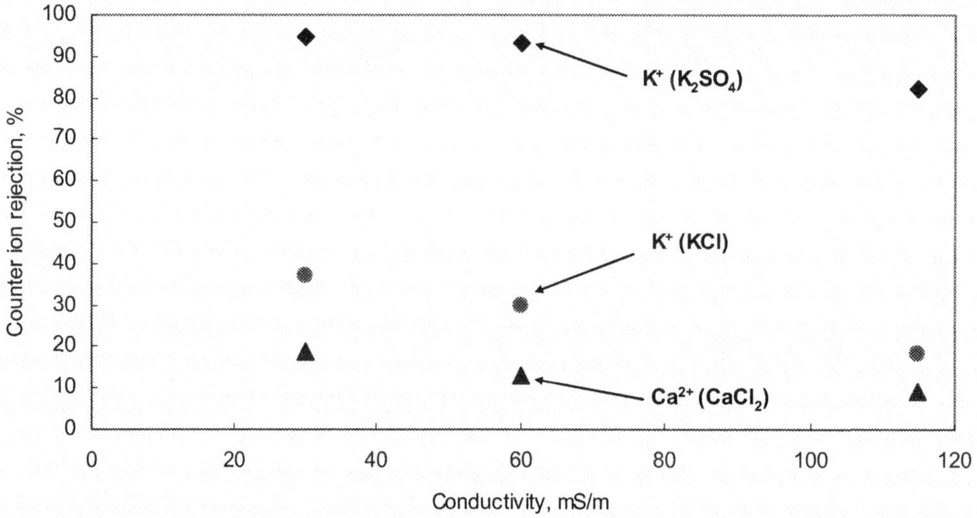

(F) Counter-ion rejection by MX07 (NF membrane)

Figure 7.7 (Continued)

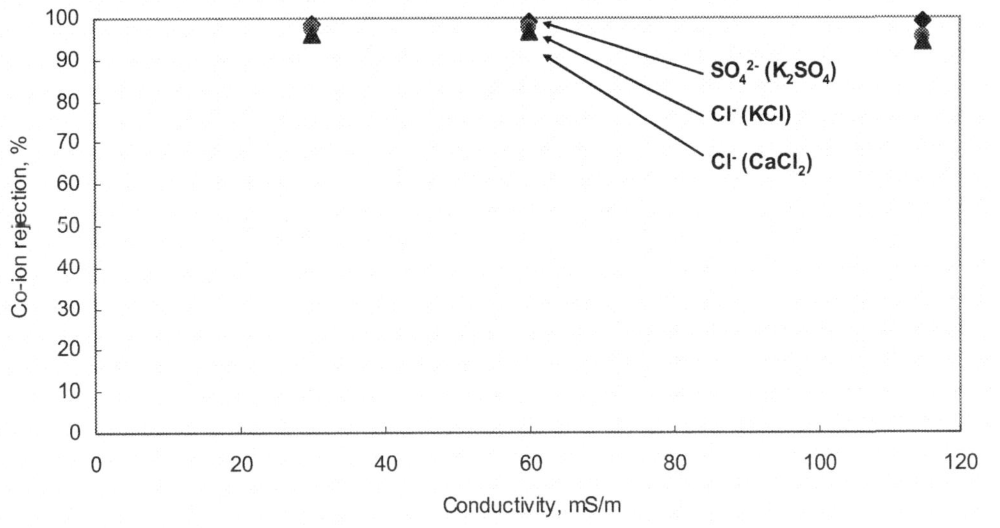

(G) Co-ion rejection by LFC-1 (RO membrane)

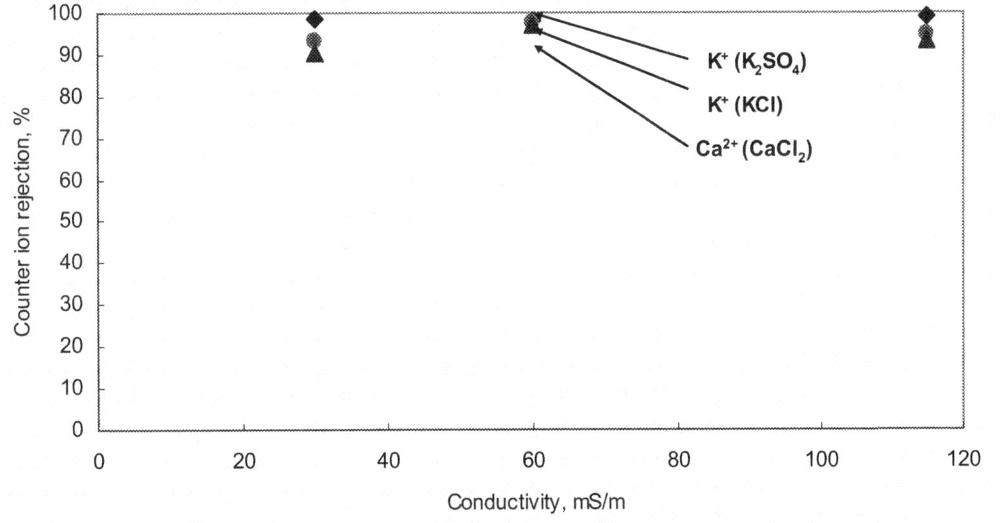

(H) Counter-ion rejection by LFC-1 (RO membrane)

Figure 7.7 (Continued)

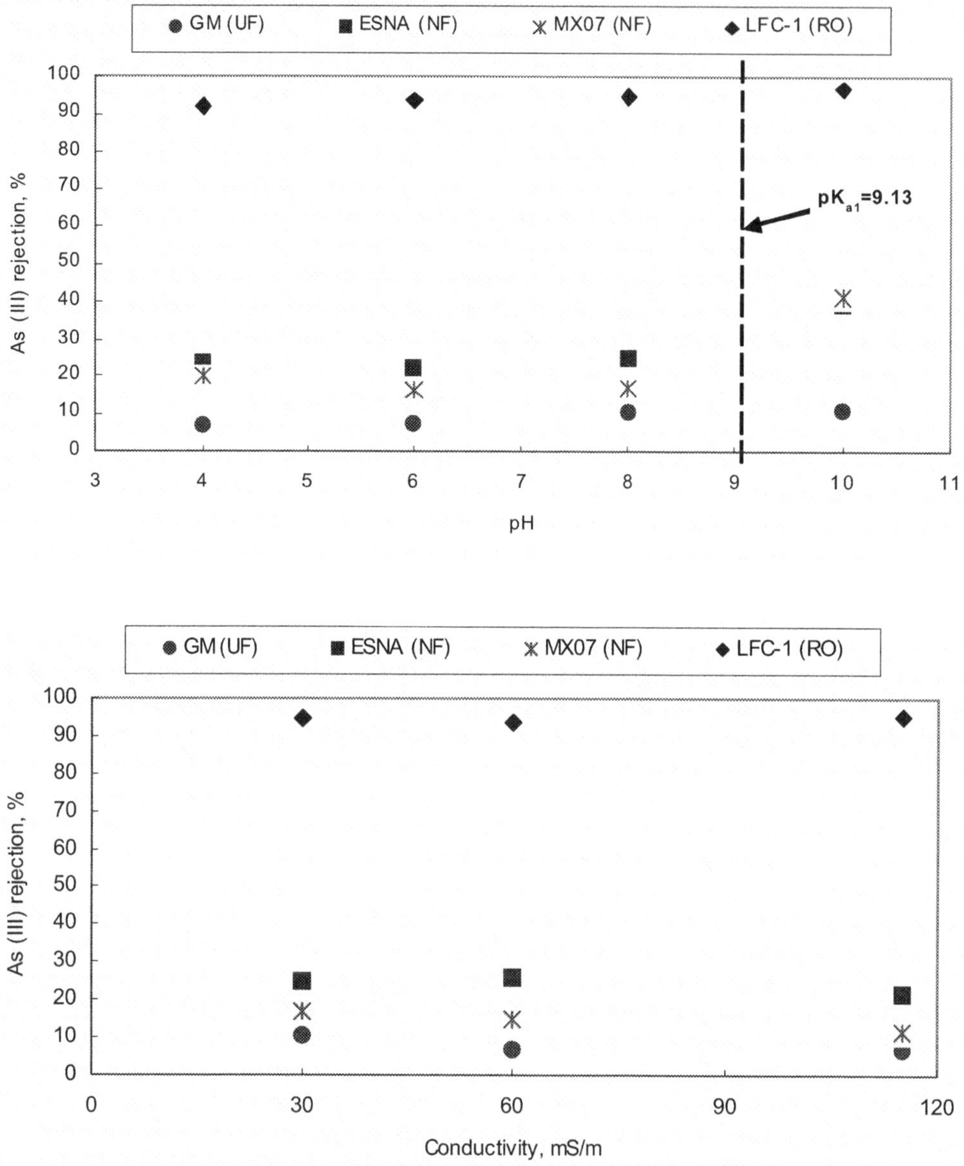

Figure 7.8 Effect of pH and conductivity on membrane rejection of As(III)

HEXAVALENT CHROMIUM REJECTION AND FLUX DECLINE IN VARIOUS NATURAL WATER SUPPLIES

Two RO membranes (LFC-1 and TFC-ULP), two NF membranes (ESNA and TFC-S), and one tight UF (GM) membrane were selected for Cr(VI) rejection tests using a cross-flow flat-sheet filtration unit. The characteristics of the five membranes are described in Chapter 3.

Three water sources were used for these tests: LADWP, Glendale, and CRW. The LADWP and Glendale sources contained 120 μg/L and 70 μg/L of Cr(VI), respectively. The CRW did not contain Cr(VI), so it was spiked with 100 μg/L.

Figures 7.9 and 7.10 show the rejection of Cr(VI) and other important parameters—dissolved organic carbon (DOC), TDS, and ultraviolet absorbance at 254 nanometers (UVA_{254})—as well as flux-decline trends for each combination of membrane and water source. Table 7.1 presents the Cr(VI) rejections of each membrane operated with the three water sources under the same conditions (2.8% recovery). The TFC-ULP is a loose (ultra-low-pressure) RO membrane whose performance was similar to that of the TFC-S, a tight NF membrane. During a 24-hr testing period, fluxes for the three water sources ranged from 62.8% to 72.8% of the initial flux with the LFC-1 and from 65.9% to 86.1% with the TFC-ULP (both RO membranes), from 74.3% to 82.6% with the TFC-S and from 75.6% to 85.2% with the ESNA (both NF membranes), and from 87.1% to 88.4% with the GM (a UF membrane).

In general, Table 7.1 and Figure 7.9 show that the tighter the membrane, the higher the percent rejection of the contaminant of interest. These results indicate that size exclusion may be the dominant mechanism for the membranes, natural water supplies, and contaminants tested in these experiments.

As indicated in Figure 7.10, source water type appeared to have some impact on flux declines over time in the five membranes tested. Flux declines with CRW were higher than those with the other two water sources for the tighter (RO) membranes. For the loose membranes (UF), there was not much change in flux over the time period studied.

As shown in Figure 7.9, rejection of Cr(VI) and other parameters by the negatively charged UF membrane was low in all three water sources. Anion rejections by the UF membrane were higher in all of the synthetic water supplies (see Figures 7.3 and 7.4) than in the natural water supplies (Figure 7.9). The trend in Figure 7.3 indicates that higher rejection is caused by decreasing conductivity. Electrostatic repulsion of the anions from the UF membrane surface was not the dominant mechanism for the natural water sources tested. It is likely that the nature and concentration of anions and cations in the natural water supplies did not significantly affect the surface charge of the UF membrane.

Table 7.1
Cr(VI) rejection with each combination of membrane and water source

	Cr(VI) rejection (%)		
Membrane	LADWP water	Glendale water	CRW
LFC-1 (RO)	98	98.8	95.5
TFC-ULP (RO)	86.3	74.2	86.3
TFC-S (NF)	75.9	82.6	85.5
ESNA (NF)	45.1	52.9	43.9
GM (UF)	3.7	8.7	1.8

MEMBRANE OXIDATIVE DAMAGE BY CR(VI) SPECIES

The goal of this part of the study was to investigate membrane damage during 12 months of exposure to a strong oxidant such as Cr(VI) at varying pH and Cr(VI) values. The commercial membranes used in the anion transport study were monitored in order to determine their tolerance against Cr(VI) effects under various experimental conditions. At the beginning of the study, it was assumed that the performance of all membranes would deteriorate because of surface damage and that changes in rejection efficiency and flux would occur as a result of long-term exposure to Cr(VI).

According to a previous study (Glater et al. 1981), the use of pretreatment chemicals such as halogens, ozone, and chlorine dioxide prior to a membrane process affected membrane performance and durability. An increase in flux and a decrease in rejection indicated chemical attack of the membrane surface. Cellulose acetate membranes were the most resistant to chemical attack; these membranes responded only to ozone and were hydrolyzed at low pH. Polyamide membranes were sensitive to ozone, halogens, and chlorine dioxide and were also hydrolyzed at low pH.

Results of Long-Term Monitoring

Flux and NaCl Rejection as a Function of CT Value

Tables 7.2 and 7.3 summarize the normalized flux (J/J_o) and NaCl rejection of the five membranes as a function of CT (concentration × contact time) at varying pH values. As shown in Tables 7.2A and 7.3A, four experimental conditions (experiments 1–4) correspond to 8.84×10^2 mg-hr/L of CT, which is relatively low—a 0.1-mg/L Cr(VI) concentration and 12 months of exposure—and pH values that varied from 4 to 10. Flux and NaCl rejection results for the five membranes were similar to those observed in the initial results.

In experiments 5–8 (Tables 7.2B and 7.3B), flux and NaCl rejection showed slight increases and decreases when the CT value increased. Experiments 5–8 correspond to 8.84×10^4 mg-hr/L of CT, which is two orders of magnitude higher than the CT in experiments 1–4. However, experiments 9–12 (Tables 7.2C and 7.3C) showed that flux and NaCl rejection increased and decreased significantly because of serious membrane damage resulting from long-term exposure to Cr(VI) with a CT three or four orders of magnitude higher than the CT values shown in Table 7.2A.

These results indicate that Cr(VI) species would not likely affect commercial membranes under conditions typical of natural water sources because results of the experiments with various CT values showed that flux and NaCl rejection were not appreciably changed. However, if high Cr(VI) concentrations are present in the source water for long periods of time (as with industrial water applications), membrane degradation—including, ultimately, membrane failure—will likely occur.

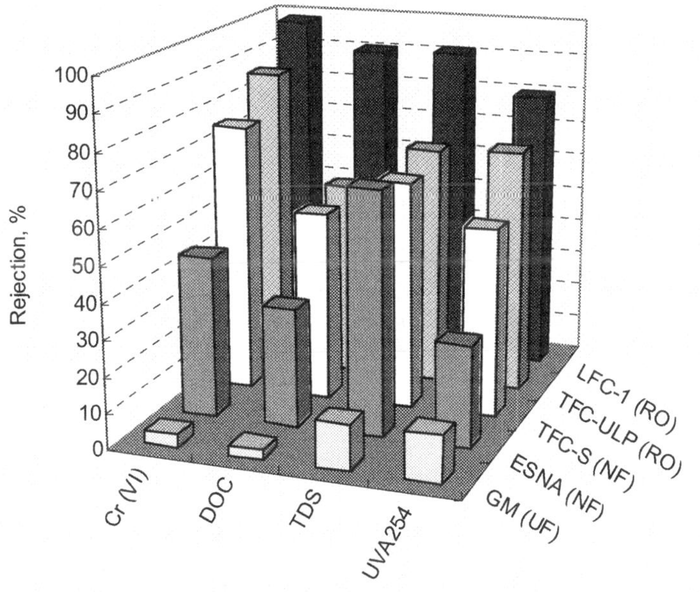

(A) LADWP water

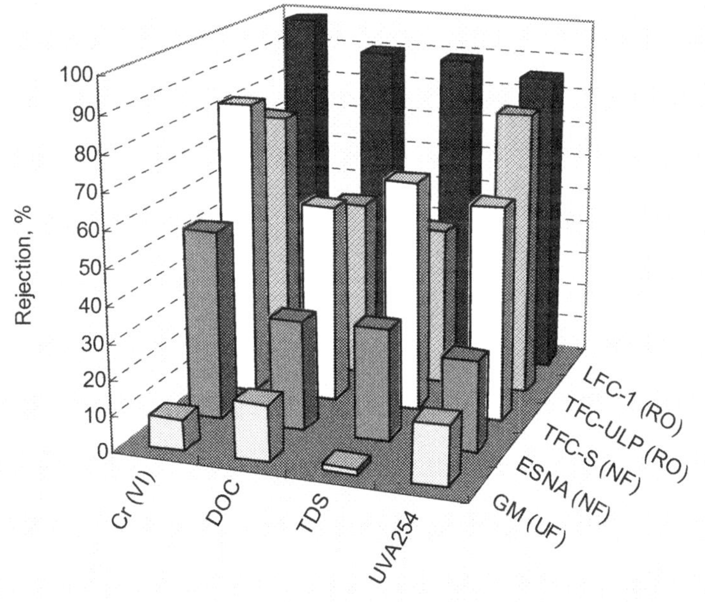

(B) Glendale water

Figure 7.9 Rejection of Cr (VI) and other parameters in natural water sources

(continued)

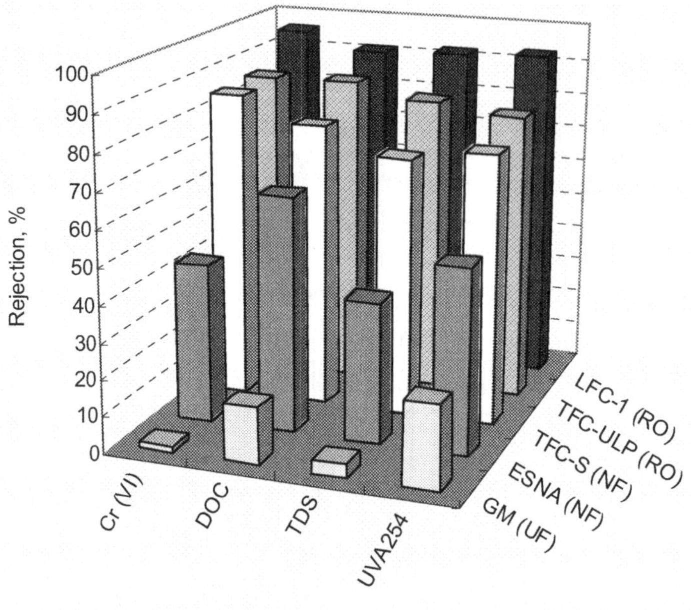

(C) Colorado River water

Figure 7.9 (Continued)

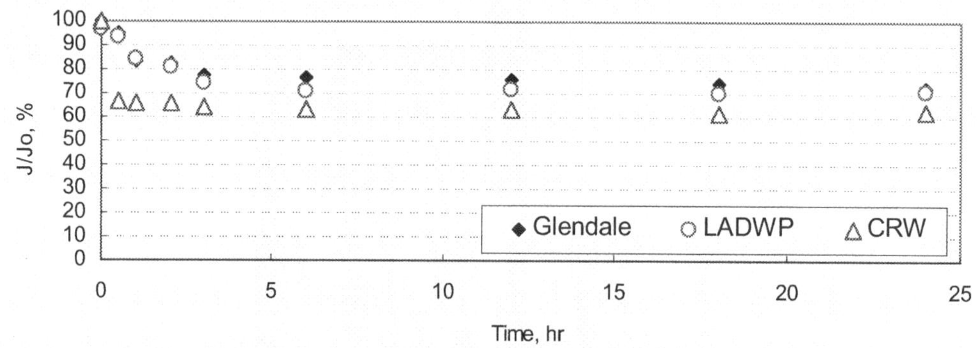

(A LFC-1 (RO membrane)

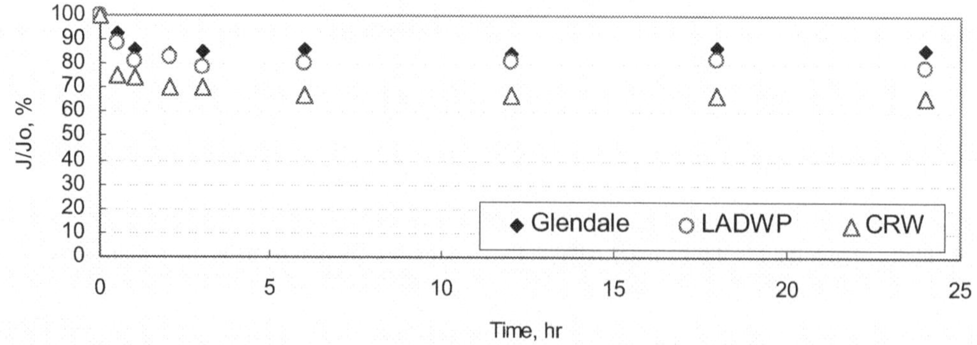

(B) TFC-ULP (RO membrane)

Figure 7.10 Flux decline in RO, NF, and tight UF membranes with natural water sources

(continued)

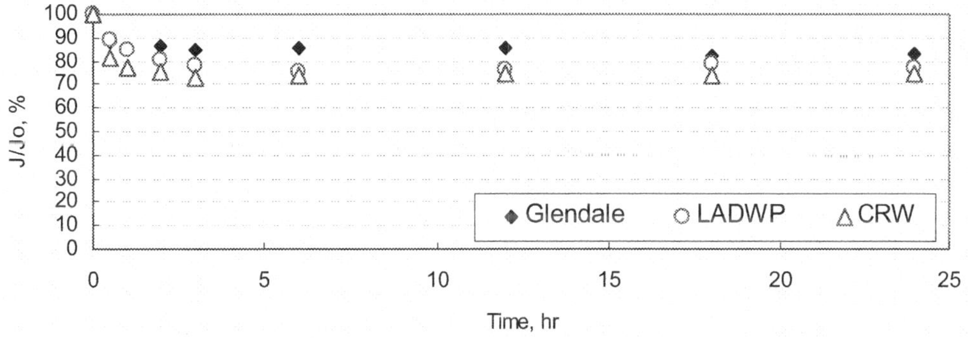

(C) TFC-S (NF membrane)

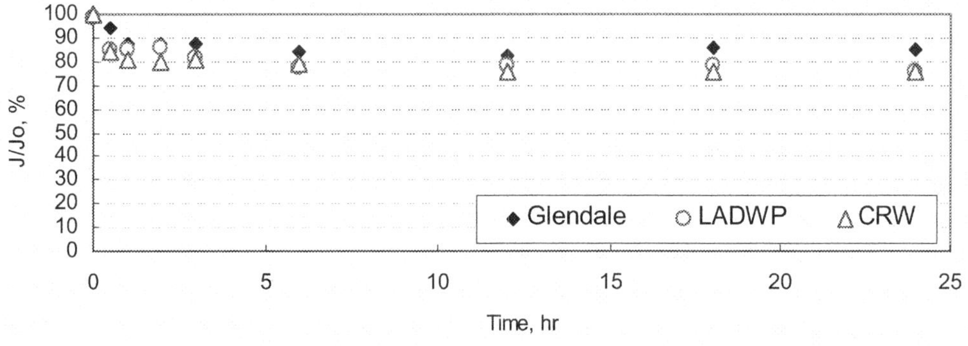

(D) ESNA (NF membrane)

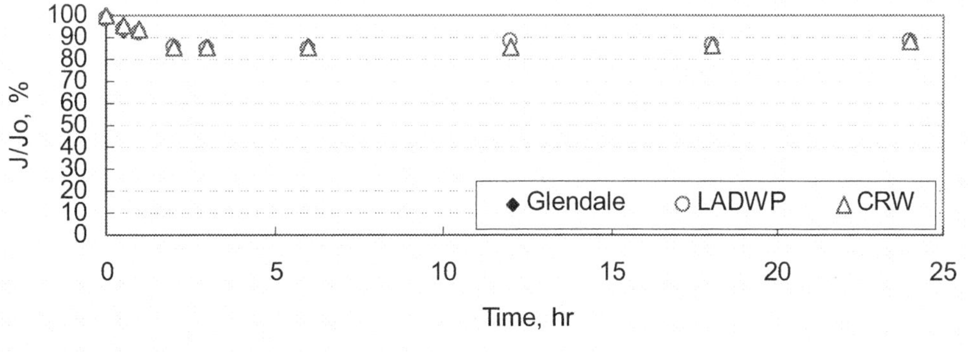

(E) GM (UF membrane)

Figure 7.10 (Continued)

Table 7.2
Effect of CT on normalized flux (J/J_o) as a function of pH (%)
(Exp 1, 5, 9: pH 4; Exp 2, 6, 10: pH = 6; Exp 3, 7, 11: pH = 8; Exp 4, 8, 12: pH ≈ 10)

(A) 0.1 mg/L Cr(VI) and varying pH

Exp No.	CT (*10^2) mg-hr/L	GM (UF)	ESNA (NF)	TFC-S (NF)	LFC-1 (RO)	TFC-ULP (RO)
1	0	100.0	100.0	100.0	100.0	100.0
1	1.08	106.3	92.0	107.2	100.0	94.2
1	2.16	103.4	90.0	93.8	99.0	97.7
1	4.32	105.6	87.1	96.7	96.7	92.9
1	8.84	105.3	87.1	94.8	96.0	88.0
2	0	100.0	100.0	100.0	100.0	100.0
2	1.08	104.8	86.0	90.0	89.0	95.2
2	2.16	104.0	92.0	89.0	95.2	91.4
2	4.32	104.8	88.0	88.0	96.0	89.5
2	8.84	105.7	85.0	85.0	95.2	84.0
3	0	100.0	100.0	100.0	100.0	100.0
3	1.08	102.3	85.0	95.0	92.0	86.2
3	2.16	100.0	87.0	91.3	93.0	89.1
3	4.32	102.3	86.0	95.0	91.0	89.1
3	8.84	104.0	85.0	95.0	90.0	87.0
4	0	100.0	100.0	100.0	100.0	100.0
4	1.08	102.0	93.0	98.0	96.0	107.4
4	2.16	108.0	96.0	106.2	98.0	112.0
4	4.32	97.0	98.0	99.5	101.8	108.5
4	8.84	102.7	93.3	100.8	93.3	107.5

(continued)

Table 7.2 (Continued)

(B) 100 mg/L Cr(VI) and varying pH

Exp No.	CT (*10^4) mg-hr/L	GM (UF)	ESNA (NF)	TFC-S (NF)	LFC-1 (RO)	TFC-ULP (RO)
5	0	100.0	100.0	100.0	100.0	100.0
5	1.08	96.0	104.8	94.3	85.0	85.5
5	2.45	101.0	104.0	94.3	88.0	86.7
5	4.32	109.5	104.0	96.0	108.6	89.5
5	8.84	108.0	100.0	100.0	111.4	92.0
6	0	100.0	100.0	100.0	100.0	100.0
6	1.08	105.6	92.0	87.6	92.8	98.5
6	2.45	107.8	85.3	94.7	96.6	98.5
6	4.32	115.0	87.2	99.5	99.5	98.5
6	8.84	117.5	104.0	98.5	104.0	89.1
7	0	100.0	100.0	100.0	100.0	100.0
7	1.08	104.2	94.2	110.6	97.1	92.8
7	2.45	107.0	95.2	102.6	98.0	93.0
7	4.32	111.2	94.2	88.9	99.0	90.0
7	8.84	113.1	97.1	89.6	100.8	100.8
8	0	100.0	100.0	100.0	100.0	100.0
8	1.08	119.1	101.3	102.4	104.0	100.0
8	2.45	133.1	96.0	94.9	105.9	98.4
8	4.32	135.0	99.4	84.7	104.0	98.4
8	8.84	132.0	110.6	91.0	104.0	98.4

(continued)

Table 7.2 (Continued)

(C) 10,000 mg/L Cr(VI) and varying pH

Exp No.	CT (*10^6) mg-hr/L	GM (UF)	ESNA (NF)	TFC-S (NF)	LFC-1 (RO)	TFC-ULP (RO)
9	0	100.0	100.0	100.0	100.0	100.0
9	1.08	133.6	88.9	100.7	113.0	122.9
9	2.45	143.1	106.8	113.5	113.0	136.5
9	4.32	146.8	104.8	107.5	131.0	145.6
9	8.84	168.0	116.4	115.9	140.0	157.2
10	0	100.0	100.0	100.0	100.0	100.0
10	1.08	110.0	102.0	97.0	112.0	92.1
10	2.45	123.0	120.0	95.2	116.0	97.1
10	4.32	134.0	116.0	101.8	118.0	97.1
10	8.84	140.0	120.0	102.7	119.0	101.8
11	0	100.0	100.0	100.0	100.0	100.0
11	1.08	99.3	87.2	94.1	94.0	92.0
11	2.45	104.0	114.0	103.0	114.0	99.9
11	4.32	119.0	123.0	106.0	119.0	100.9
11	8.84	120.0	130.0	99.8	130.0	101.6
12	0	100.0	100.0	100.0	100.0	100.0
12	1.08	137.2	80.1	87.8	92.0	83.5
12	2.45	153.0	108.0	98.3	106.0	102.3
12	4.32	157.0	124.0	99.0	106.0	107.7
12	8.84	162.6	140.0	101.6	120.0	109.7

Table 7.3
Effect of CT on normalized NaCl rejection as a function of pH (%)
(Exp 1, 5, 9: pH 4; Exp 2, 6, 10: pH = 6; Exp 3, 7, 11: pH = 8; Exp 4, 8, 12: pH ≈ 10)

(A) 0.1 mg/L Cr(VI) and varying pH

Exp No.	CT ($*10^2$) mg-hr/L	GM (UF)	ESNA (NF)	TFC-S (NF)	LFC-1 (RO)	TFC-ULP (RO)
1	0	100	100.0	100.0	100.0	100.0
1	2.16	101.1	97.0	95.4	90.0	91.7
1	4.32	90.4	84.8	96.2	92.5	86.7
1	8.84	94	89.1	94.1	87.5	83.3
2	0	100.0	100.0	100.0	100.0	100.0
2	2.16	89.1	83.7	101.7	92.7	98.4
2	4.32	86.7	83.2	105.6	92.0	104.8
2	8.84	85.0	82.0	102.6	90.9	106.5
3	0	100.0	100.0	100.0	100.0	100.0
3	2.16	86.0	99.7	89.8	91.6	92.2
3	4.32	83.6	89.0	93.4	89.2	93.2
3	8.84	80.6	79.4	90.4	84.3	92.6
4	0	100.0	100.0	100.0	100.0	100.0
4	2.16	103.2	101.0	98.5	94.8	95.1
4	4.32	104.4	92.0	101.4	94.8	94.8
4	8.84	100.0	89.4	96.8	93.9	87.6

(continued)

Table 7.3 (Continued)

(B) 100 mg/L Cr(VI) and varying pH

Exp No.	CT ($*10^4$) mg-hr/L	GM (UF)	ESNA (NF)	TFC-S (NF)	LFC-1 (RO)	TFC-ULP (RO)
5	0	100.0	100.0	100.0	100.0	100.0
5	2.16	100.9	93.2	98.2	99.4	96.1
5	4.32	98.9	75.4	89.2	94.7	84.4
5	8.84	79.6	71.2	90.0	86.8	84.0
6	0	100.0	100.0	100.0	100.0	100.0
6	2.16	90.9	84.5	99.1	95.0	92.2
6	4.32	92.0	86.3	85.7	85.3	73.4
6	8.84	50.9	65.2	74.6	45.3	59.2
7	0	100.0	100.0	100.0	100.0	100.0
7	2.16	83.3	75.5	84.9	78.4	83.1
7	4.32	79.2	49.9	70.3	58.9	72.4
7	8.84	42.5	51.6	78.0	69.4	65.8
8	0	100.0	100.0	100.0	100.0	100.0
8	2.16	100.0	92.9	92.2	75.9	82.6
8	4.32	96.7	84.8	89.6	68.6	71.8
8	8.84	53.3	87.9	73.3	47.8	64.2

(continued)

Table 7.3 (Continued)

(C) 10,000 mg/L Cr(VI) and varying pH

Exp No.	CT (*10^6) mg-hr/L	GM (UF)	ESNA (NF)	TFC-S (NF)	LFC-1 (RO)	TFC-ULP (RO)
9	0	100	100.0	100.0	100.0	100.0
9	2.16	71.2	78.7	85.4	81.5	81.2
9	4.32	56	26.2	56.0	61.9	53.1
9	8.84	36.9	17.1	46.0	61.1	30.1
10	0	100.0	100.0	100.0	100.0	100.0
10	2.16	75.4	87.3	92.9	70.6	78.0
10	4.32	58.5	70.7	66.5	52.4	58.3
10	8.84	50.8	40.5	46.8	17.3	33.6
11	0	100.0	100.0	100.0	100.0	100.0
11	2.16	84.5	89.5	84.3	94.6	94.8
11	4.32	52.4	69.3	79.4	78.5	95.0
11	8.84	19.4	65.9	54.2	47.1	47.8
12	0	100.0	100.0	100.0	100.0	100.0
12	2.16	70.3	77.1	88.6	96.2	96.1
12	4.32	51.4	52.4	83.4	78.3	91.8
12	8.84	47.0	49.5	82.8	54.1	72.9

Comparison of Clean and Damaged Membranes by FTIR Analysis

Figure 7.11 presents the FTIR spectra of a clean GM membrane and a GM membrane damaged by a high Cr(VI) concentration (CT of 8.84×10^6 mg-hr/L). Infrared spectroscopy has been widely used to identify the functional groups of membranes or adsorbed organic and inorganic compounds on the membrane surface (Her, Amy, and Jarusutthirak 2000; Her 2002).

As shown in Figure 7.11, characteristic infrared peaks of polysulfone were observed at 1,592 cm^{-1} and 1,100 cm^{-1} (aromatic double-bonded carbons), at 1,016 cm^{-1} (ether), at 1,492 cm^{-1} (methyl), and at 1,151 cm^{-1} and 694 cm^{-1} (sulfone) (Bellamy 1975). In addition, the membrane showed infrared peaks at 1,650 cm^{-1}, indicating an amide group. This amide peak was not present with a CT of 8.84×10^6 mg-hr/L (experiment 9 at pH 4), indicating that one important functional group, amide, was missing from the GM membrane as a result of degradation by Cr(VI). This result is consistent with the results of an earlier study conducted by Beverly, Seal, and Hong 2000).

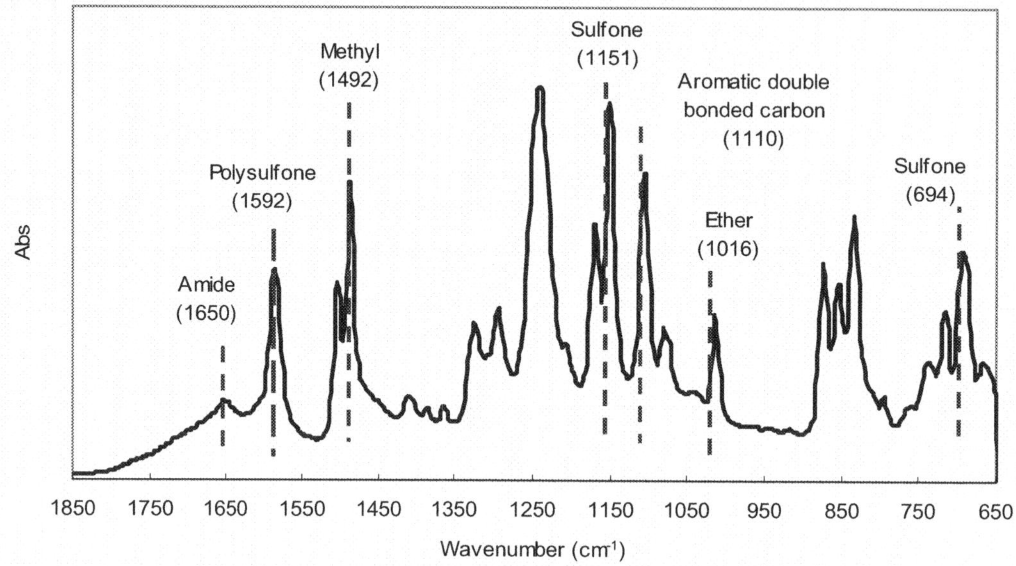

(A) Clean membrane

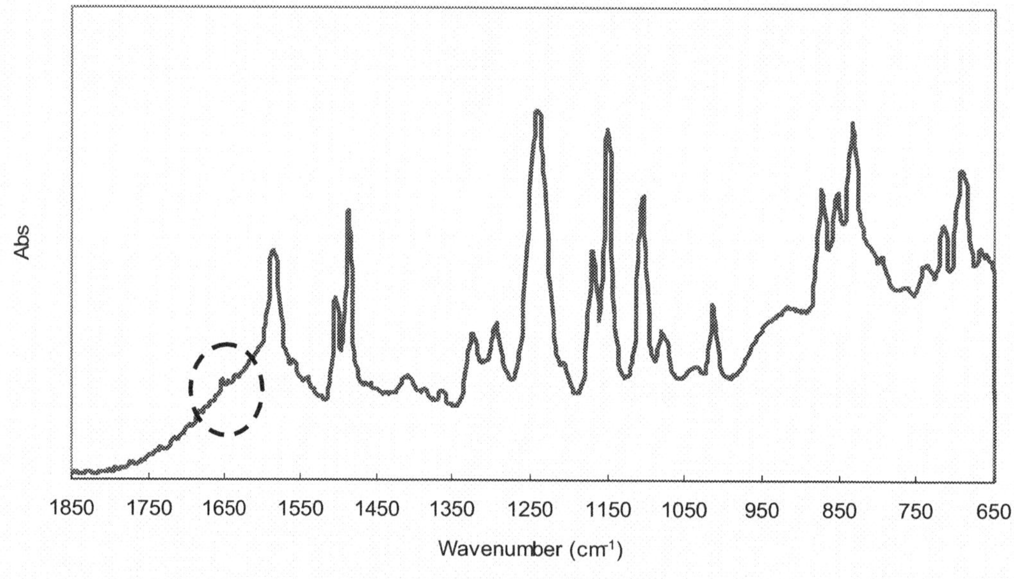

(B) Damaged membrane

Figure 7.11 FTIR spectra of clean and damaged GM membranes (pH 4, CT = 8.84×10^6 mg-hr/L

SUMMARY

In studies of synthetic water, NF and UF rejections of three anions increased with increasing pH at the same conductivity (30 mS/m), regardless of electrolyte type. Under the same pH and conductivity conditions, NF and UF rejection efficiencies for the target anions followed the order $K_2SO_4 \geq KCl > CaCl_2$. However, anion rejections by the RO membranes were higher under all the experimental conditions. Anion rejections followed the order RO > NF > UF, according to the dominant rejection mechanisms hypothesized. These results support the research hypothesis that anion rejection is enhanced by the negative surface charge (membrane zeta potential) caused by increasing pH and by lower MMCO.

The membranes rejected co- and counter-ions—SO_4^{2-}, Cl^-, K^+, and Ca^{2+}—depending on pH and conductivity (ionic strength). The divalent co-ion SO_4^{2-} was rejected more than the monovalent co-ion Cl^- in the presence of K^+ at varying pH and conductivity. The counter-ion K^+ was rejected more in the presence of SO_4^{2-} than in the presence of Cl^-. These ion rejections are also related to zeta potential. These results support the research hypothesis that higher conductivity can reduce anion rejections because the membrane surface charge is made less negative. The negative charge of the membrane surface was more reduced in the presence of the divalent counter-ion Ca^{2+} than in the presence of the monovalent counter-ion K^+ because an increase in concentration–polarization in the presence of Ca^{2+} at the membrane surface resulted in the surface charge becoming significantly less negative.

As(III) rejection was tested to examine electrostatic interactions with negatively charged membranes. As(III) rejections varied slightly at pH values of 4, 6, and 8 and constant conductivity (30 mS/m, adjusted with KCl) because the As(III) exists as an uncharged species below pH 9.13 (pKa=9.13). However, As(III) rejection increased at pH 10, at which As(III) is changed to an anionic species. These results support the premise that electrostatic interaction (repulsion) occurs between the anion species and the negatively charged membranes.

Rejections of the target anions were higher in synthetic water sources than in natural water supplies. In general, the membrane rejection data show that for natural water supplies, the tighter the membrane, the higher the percent rejection of the contaminant of interest. These results indicate that size exclusion may be the dominant mechanism for the membranes, natural water supplies, and contaminants tested in these experiments.

Study results indicate that membranes were damaged by exposure to very high concentrations of the oxidant Cr(VI). Changes in flux and NaCl rejection demonstrated that the membranes were damaged under specific conditions, and FTIR analysis confirmed membrane damage.

Nevertheless, the results indicate that the commercial membranes LFC-1, TFC-ULP, ESNA, TFC-S, and GM may not be affected by Cr(VI) species under conditions typical of natural water supplies. Considering *CT* values and membrane average life, and based on changes in flux and NaCl rejection, membranes will be not degraded by normal Cr(VI) concentrations in natural water supplies. However, if Cr(VI) concentrations are uncharacteristically high and membrane treatment is required for a long period, the possibility of membrane failure should be considered.

High-pressure membranes can be used to reject Cr(VI) and other toxic anions. The key choice is between NF membranes, which provide higher specific flux and more solute selectivity, and RO membranes, which provide greater solute removal (≥90%). If NF is selected, the negative surface charge of the membrane is an important attribute. Water quality conditions (pH and ionic strength) play an important role in membrane selection. NF is more viable under conditions of higher pH and lower ionic strength and in situations in which NF rejection can approach that of RO.

CHAPTER 8
ASSESSMENT OF HEXAVALENT CHROMIUM TREATMENT BY IRON-INDUCED REDUCTION AND PRECIPITATION–COPRECIPITATION AND COAGULATION

BACKGROUND

The removal of Cr(VI) by reduction with ferrous iron [Fe(II)] and subsequent precipitation, coprecipitation, or coagulation with ferric iron [Fe(III)] has long been used in industrial chromium treatment processes. The process takes advantage of Cr(VI)'s relatively strong properties as an oxidant, Cr(III)'s relatively low solubility as chromium hydroxide at moderate pH, and the ability of iron hydroxides to coagulate Cr(III) precipitates. The chromium reduction and precipitation process is illustrated in Figure 8.1.

Chemically, the process involves an oxidation–reduction reaction in which chromium and iron exchange electrons, thereby changing their oxidation states. Ferrous iron acts as a reducing agent, supplying the electrons for the reaction. Cr(VI) is the oxidizing agent and serves as a sink for electrons produced by ferrous iron oxidation. The pertinent half reactions are

$$Fe^{2+} \Rightarrow Fe^{3+} + e^- \qquad E^0 = -0.771\ V \qquad (8.1)$$

$$Cr^{6+} + 3e^- \Rightarrow Cr^{3+} \qquad E^0 = 1.19\ V \qquad (8.2)$$

An alternative reduction process could use elemental iron as the source of electrons, with a half reaction as follows:

$$Fe^0 \Rightarrow Fe^{2+} + 2e^- \qquad E^0 = 0.441\ V \qquad (8.3)$$

The full oxidation–reduction reaction using ferrous iron as a reductant is

$$3\ Fe(OH)_2 + CrO_4^{2-} + 4\ H_2O \Rightarrow 3\ Fe(OH)_3 + Cr(OH)_3 + 2\ OH^- \qquad (8.4)$$

As shown in equation 8.4, the stoichiometric ratio for the redox reaction is 3 moles of iron to 1 mole of chromium. Conveniently, the molecular weights of iron and chromium are approximately equal, 55.8 and 52 respectively, so the stoichiometric weight ratio is also approximately 3:1.

If the reaction occurs at a moderate pH, $Cr(OH)_3$ (Ksp 6.3×10^{-31}) will precipitate and can be removed from solution. In a practical treatment scheme, the mass of iron required to complete the reduction of Cr(VI) will exceed the stoichiometric amount in order to increase the speed of the reaction and to satisfy other reductant demands. If the intent is to coagulate Cr(III) precipitates, additional excess iron can be added as ferrous or ferric iron to form iron floc. When iron is added in the ferrous state, adequate oxidizing agents must be present to convert the ferrous to ferric because $Fe(OH)_2$ is much more soluble than $Fe(OH)_3$ (Ksp = 8×10^{-16} and 4×10^{-38} respectively). Typically, dissolved oxygen can serve as the sink for the surplus electrons needed to complete the iron oxidation reaction.

The five phases (A through E) evaluated in these experiments are described in Chapter 3.

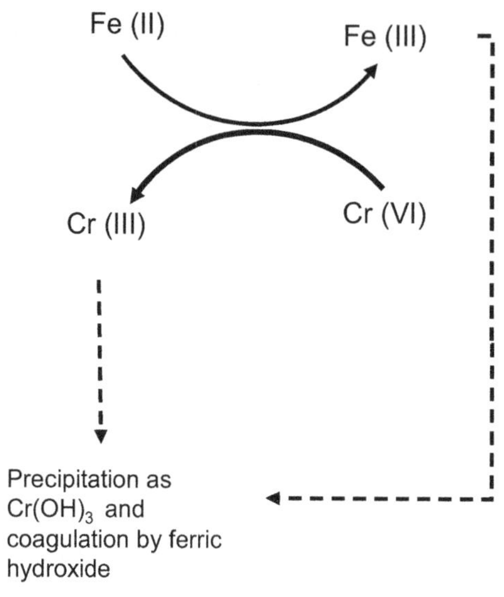

Figure 8.1 The chromium reduction and precipitation/coagulation process

PHASE A—EVALUATION OF IRON REDUCTION KINETICS

The kinetic performance of Cr(VI) reduction by Fe(0) and Fe(II) was compared as a function of reductant dose and pH. The iron reductant was added at Fe:Cr weight ratios of 3:1, 6:1, and 10:1, or approximately one, two, or three times the stoichiometric ratio needed for the Fe(II) reaction. For Fe(0) reduction only, the experiments were performed at pH 7.5, 6.5, 5.5, and 4.5. After the reaction was initiated, samples were taken periodically over a 30- or 120-min period. The experiments were performed with synthetic water, which had an alkalinity of 50 mg/L as $CaCO_3$ and 0.1 M KCl to control ionic strength. The synthetic water was spiked to achieve an initial Cr(VI) concentration of 100 μg/L. The experiments were performed open to the atmosphere and at room temperature (≈20°C).

Figure 8.2 presents the results of these experiments. Fe(II) was an effective reductant. Kinetically, the Fe(II) reduction was relatively fast, with the bulk of reduction occurring within 15 min. At all doses investigated, the rate and extent of reduction improved as the pH was increased from 5.5 to 7.5. Increasing the Fe(II):Cr(VI) weight ratio also increased the rate and extent of Cr(VI) reduction. In this water matrix, 95% reduction of Cr(VI) to Cr(III) could be accomplished within 10 min at pH 7.5 and an Fe(II):Cr(VI) weight ratio of 6:1.

Under the conditions tested, Fe(0) was not an effective reductant. At best, a 10–15% reduction of Cr(VI) to Cr(III) was obtained. No clear trend with respect to the influence of pH was observed, and increasing the Fe(II):Cr(VI) weight ratio did not appear to improve Cr(VI) reduction.

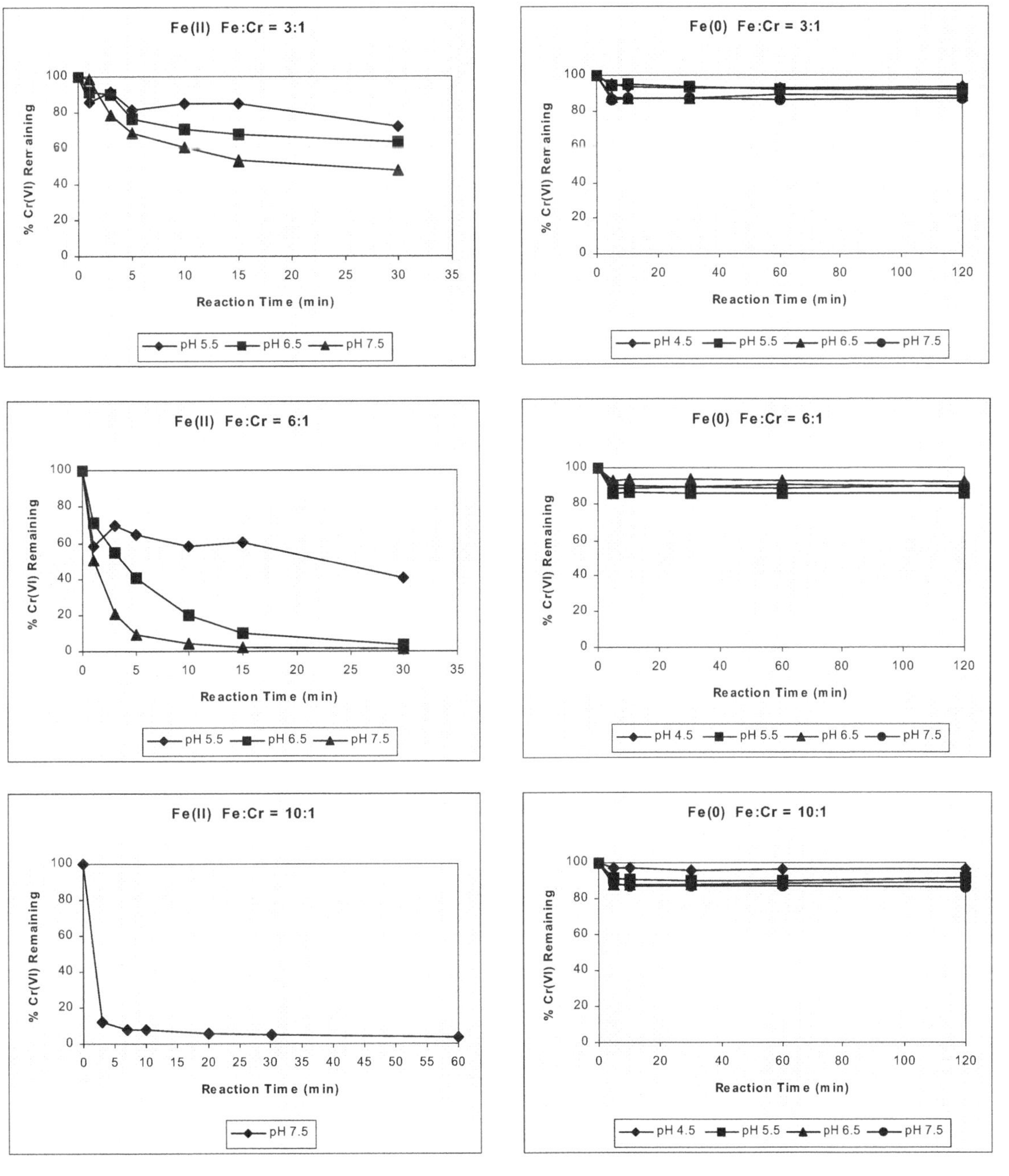

Figure 8.2 Comparison of Cr(VI) reduction by Fe(II) and Fe(0) in synthetic water

PHASE B—IMPACT OF CO-OCCURRING PARAMETERS ON REDUCTION KINETICS

The kinetic performance of Cr(VI) reduction by Fe(0) and Fe(II) in the presence of co-occurring contaminants was compared as a function of increasing co-occurring contaminant concentration. The co-occurring contaminants studied were phosphate, sulfate, arsenate, and silica. Experiments were conducted with synthetic water spiked with individual co-occurring contaminants at concentrations of 0.1, 0.25, and 0.4 mg/L. The iron reductant was added at a weight ratio of 10:1 Fe:Cr, and the experiments were performed at a constant pH of 7.5. The synthetic water, which had an alkalinity of 50 mg/L as $CaCO_3$ and 0.1 M KCl to control ionic strength, was spiked to an initial Cr(VI) concentration of 100 μg/L. The experiments were conducted open to the atmosphere and at room temperature. After the reaction was initiated, samples were taken periodically over a 1,440-min (one-day) period.

Figure 8.3 presents the results of these experiments. Once again, Fe(II) was a far more effective reductant than Fe(0). In the absence of co-occurring contaminants nearly complete Cr(VI) reduction by Fe(II) was observed. On the other hand Cr(VI) reduction by Fe(0) was poor; only about 20% of the Cr(VI) was reduced to Cr(III) over a 1,440-min period. Because of the poor performance of Fe(0), no additional zero-valent iron experiments were conducted in experimental Phases C and D.

The presence of phosphate, sulfate, arsenate, or silica had varying effects on the reduction of Cr(VI) by Fe(II). Figure 8.3 shows that, at the concentrations and pH tested, the presence of sulfate had no impact on Cr(VI) reduction. The effects of arsenate and phosphate were comparable to one another. In both cases, increasing concentrations of either phosphate or arsenate appeared to slightly decrease the amount of Cr(VI) that was reduced to Cr(III). In the absence of either contaminant, Cr(VI) reduction was essentially complete, whereas approximately 15% of the Cr(VI) remained unreduced after 1,440 min in the presence of 0.4 mg/L arsenate or phosphate.

The presence of even a slight amount of silica appeared to inhibit Cr(VI) reduction at pH 7.5. The addition of 0.1 mg/L silica as Si significantly decreased the amount of Cr(VI) that was reduced to Cr(III). At this relatively low silica dose (compared with silica concentrations in natural water sources), only 40% of the Cr(VI) was reduced to Cr(III). When the silica spike was increased to 0.4 mg/L, the amount of Cr(VI) reduced to Cr(III) decreased to approximately 30%. Because silica is unlikely to participate directly in any oxidation–reduction reactions, the mechanism by which silica interferes with Cr(VI) reduction is unclear. However, later in this chapter a possible mechanism accounting for this effect will be proposed.

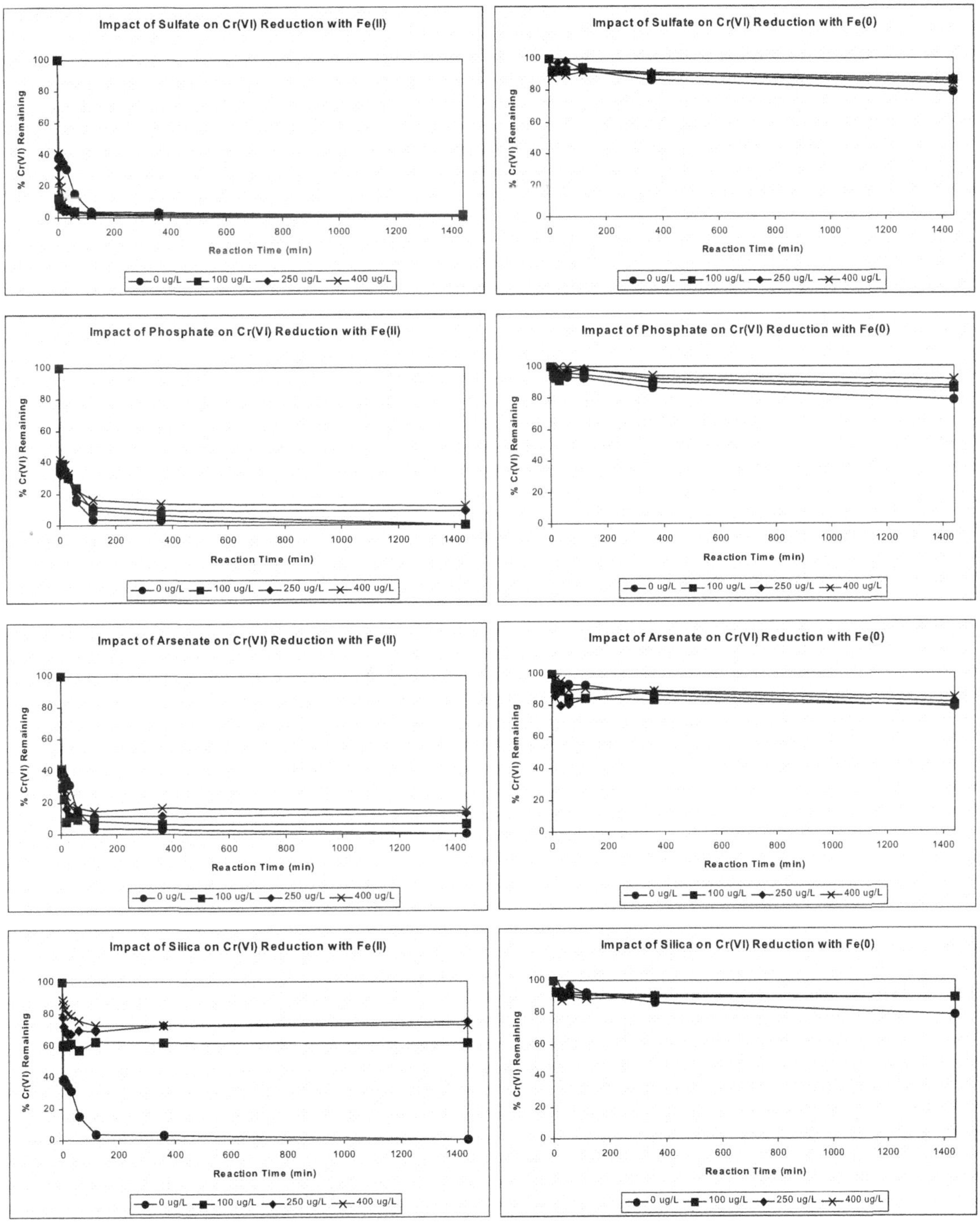

Figure 8.3 Impact of co-occurring contaminants sulfate, phosphate, arsenate, and silica on Cr(VI) reduction by Fe(II) and Fe(0)

PHASE C—TOTAL CHROMIUM REMOVAL IN SYNTHETIC WATER

Removal of total chromium from solution after the reduction of Cr(VI) to Cr(III) was simulated by passing aliquots of the Fe(II)-treated solutions through membrane filters of various pore sizes. Like those in Phases A and B, experiments in Phase C were conducted with synthetic water with an alkalinity of 50 mg/L as $CaCO_3$ and 0.1 M KCl to control ionic strength. The synthetic water was spiked to an initial Cr(VI) concentration of 100 µg/L, and pH was adjusted to 7.5. The experiments were performed open to the atmosphere and at room temperature.

The extent of chromium removal by filtration after reduction with Fe(II) was analyzed as a function of reductant dose, time, and filter pore size. The Fe(II) reductant was added at Fe:Cr weight ratios of 10:1, 50:1, and 100:1. After the reaction was initiated, samples were taken periodically over a 1,440-min period and passed through the membrane filter. The size range of the resulting precipitates was inferred by determining whether chromium passed through the filter. The membranes used to filter the Fe(II)-treated solution ranged from a 0.45-µm microfilter to a 10,000-dalton ultrafilter. Because there is no analytical technique capable of directly measuring Cr(III), the ability of the filter to remove chromium from the solution was estimated by measuring total chromium—the sum of Cr(VI) and Cr(III) concentrations. Because the experiments were run at conditions under which Fe(II) almost totally reduced Cr(VI) to Cr(III), the total chromium measurement provided a reasonable estimate of Cr(III) concentration.

Results of this simulation of the reduction–precipitation treatment process are presented in Figure 8.4. As shown in the figure, the 10:1 Fe(II) dose reduced essentially all of the Cr(VI) in the synthetic water to Cr(III). Still, a substantial portion of total chromium was not removed by the 0.45-µm and 0.2-µm filters, indicating that the size of some precipitates was smaller than the pore sizes of the filters. Approximately 40% of the total chromium—presumably all Cr(III)—passed through the 0.45-µm filter and 20% through the 0.2-µm filter. Thus, ferrous iron was capable of reducing the Cr(VI) to Cr(III), but the precipitates and coprecipitates formed by this process were too small to be removed by the 0.45-µm and 0.2-µm filters. In fact, assuming the operational definition of dissolved species as being able to pass through a 0.45-µm filter, approximately 40% of the Cr(III) was "dissolved." This implies that substantially more iron is required to coprecipitate or coagulate Cr(III), that very small precipitates or flocs are formed, or that Cr(III) solubility limits the ability of Fe(II) to reduce chromium in solution to very low levels.

A second set of experiments was performed to fractionate the Fe(II)-treated water with filters of decreasing pore size. In addition to the 0.45-µm and 0.2-µm microfilters, ultrafilters with pore sizes of 0.02 µm, 100,000 daltons, and 10,000 daltons were used to filter the Fe(II)-treated water. The Fe:Cr weight ratio was also increased from 10:1 to 50:1 and 100:1. The results of these experiments are summarized in Figure 8.5. With the exception of the 10,000-dalton filter, approximately 20–30% of the initial total chromium concentration passed through the filters. Increasing the Fe:Cr weight ratio from 10:1 to 50:1 slightly increased the amount of total chromium retained by all the filters except for the 10,000-dalton filter. An increase from 50:1 to 100:1 did not appear to increase the quantity of total chromium retained by the any filter. Even at an Fe:Cr dosage of 100:1, approximately 8% of the total chromium in the system passed through a 10,000-dalton filter. In other words, even at an Fe:Cr weight ratio of 100:1, a small amount of total chromium, most likely speciated as Cr(III), remained in solution either as colloidal or dissolved chromium.

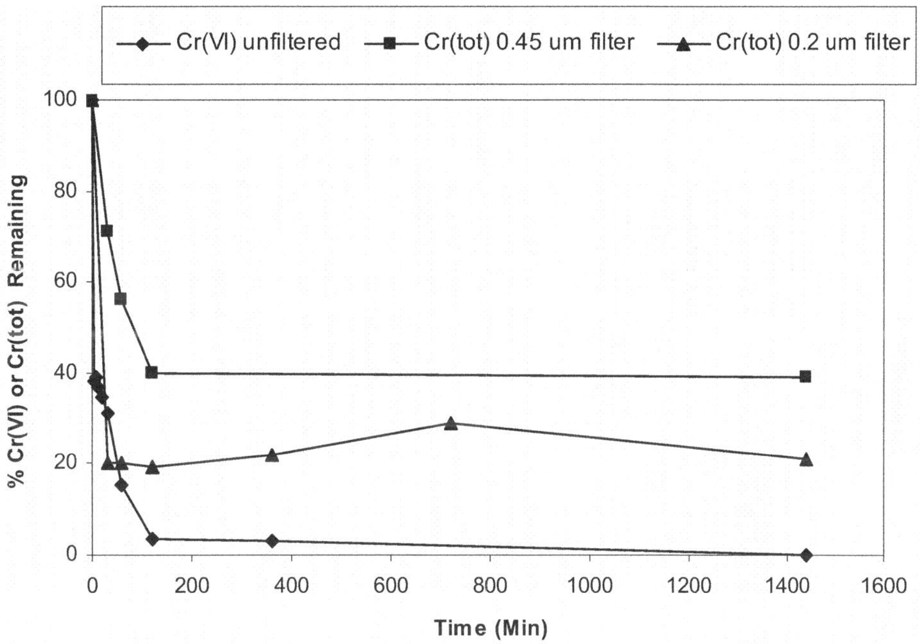

Figure 8.4 Comparison of total chromium removal in synthetic water by 0.45-µm and 0.2-µm filters at an Fe(II):Cr(VI) dosage of 10:1

PHASE D—REDUCTION KINETICS AND TOTAL CHROMIUM REMOVAL IN EXPERIMENTS WITH A NATURAL WATER SOURCE SUPPLIED BY LADWP

The kinetic performance of Cr(VI) reduction by Fe(II) in a natural water source supplied by LADWP was evaluated as a function of reductant dose and pH. The Fe(II) reductant was added at Fe:Cr weight ratios of 3:1, 6:1, and 10:1, approximately one, two, or three times the stoichiometric dose. The experiments were conducted at pH 8.0 (ambient), 7.5, and 6.5. After the reaction was initiated, samples were taken periodically over a 1,440-min period and passed through a 0.2-µm membrane filter. Both Cr(VI) and total chromium were measured after filtration.

The water used for the Phase D experiments was drawn from LADWP's AT002 well that was out of service and not supplying water to consumers. Water quality characteristics for this well are given in Table 8.1. As received, the water contained 45 µg/L total chromium and 2 µg/L Cr(VI), indicating that the chromium in this sample was primarily speciated as Cr(III). Before the experiments were run, the water was spiked to an initial Cr(VI) concentration of 100 µg/L. Thus, the total chromium concentration of the spiked AT002 water was approximately 145 µg/L. All experiments were performed open to the atmosphere and at room temperature.

Figure 8.6 presents the results of the experiments with water from well AT002. As in the experiments with synthetic water, Fe(II) effectively reduced Cr(VI) to Cr(III). However the process appeared to be kinetically slower, with the reduction being completed within 120 min rather than 30 min, as measured in the synthetic water. Also similar to the experiments with synthetic water, increasing the Fe(II):Cr weight ratio in the LADWP water increased the extent of Cr(VI) reduction.

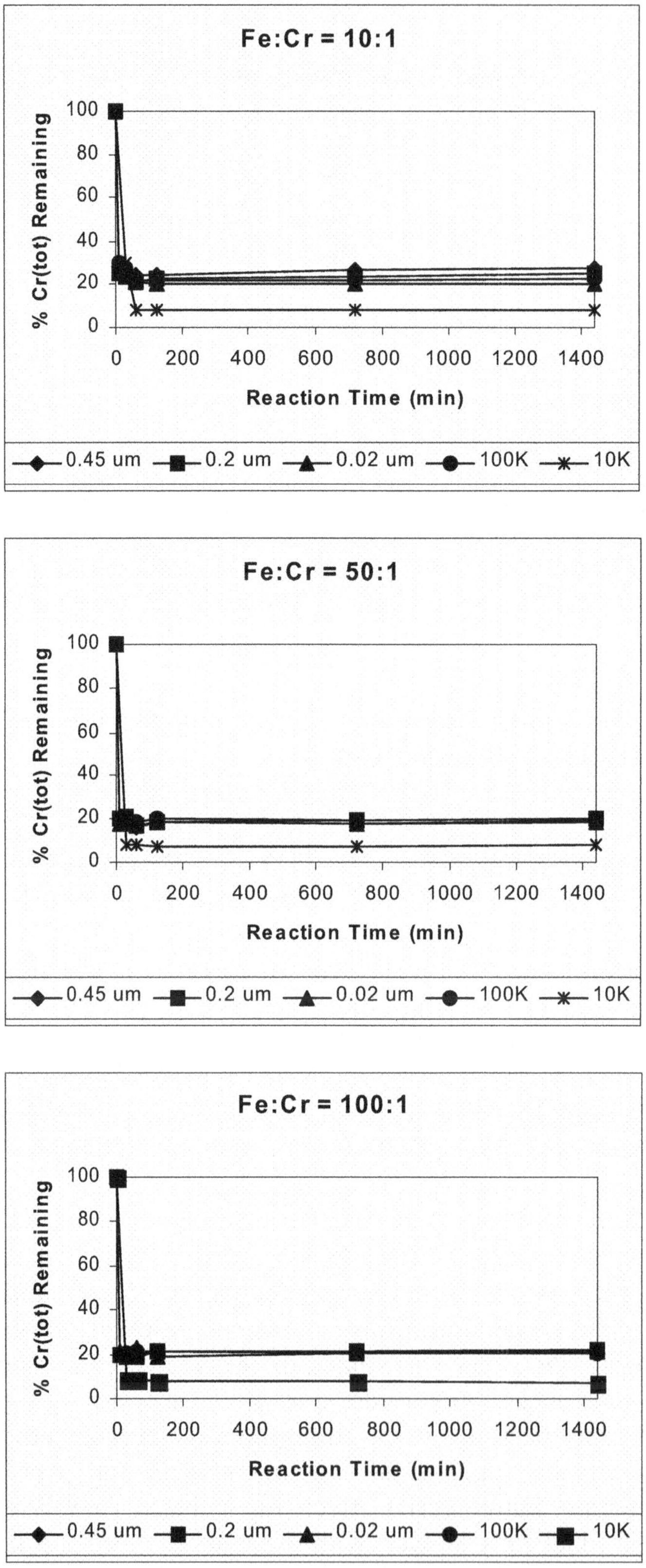

Figure 8.5 Comparison of total chromium removal in synthetic water with various filter sizes and Fe(II):Cr(VI) weight ratios of 10:1, 50:1, and 100:1

Table 8.1
Water quality characteristics of LADWP well AT002

Parameter	Unit	Value	Parameter	Unit	Value
pH	-	8.0	Mg^{2+}	mg/L	20.9
DOC	mg/L	0.67	Na^{+}	mg/L	28.7
TDS	mg/L	359	K^{+}	mg/L	4.8
Cr(VI)	µg/L	2	Cl^{-}	mg/L	34
Total chromium	µg/L	45	SO_4^{2-}	mg/L	67.8
SiO_2	mg/L	11.2	NO^{3-}	mg/L	45.9
Ca^{2+}	mg/L	95.4	PO_4^{3-}	mg/L	0

The influence of pH on Cr(VI) reduction by Fe(II) was completely different in the AT002 water compared with the synthetic water. Whereas increasing the pH from 5.5 to 7.5 improved Cr(VI) reduction in the synthetic water, decreasing the pH from 8.0 to 6.5 improved Cr(VI) reduction in the LADWP water. This trend was observed in the data measuring the percentage of Cr(VI) reduction by Fe(II) as well as the data measuring removal of total chromium by a 0.2-µm filter.

The influence of pH on the reduction–precipitation process in the LADWP water can be more clearly illustrated by comparing the chromium speciation data as a function of pH. Figure 8.7 presents these data for the experiment using an Fe:Cr weight ratio of 10:1. At pH 8.0, approximately 60% of the total chromium passing through the 0.2-µm filter was speciated as Cr(VI). At pH 7.5, approximately 40% of the total chromium passing through the 0.2-µm filter was speciated as Cr(VI). When the pH was lowered to 6.5, essentially no Cr(VI) was present, and any chromium passing through the filter was speciated as Cr(III). The amount of Cr(III) that passed through the filter also decreased with decreasing pH.

Because the LADWP water contained both Cr(III) and spiked Cr(VI), three measures of performance are significant. The first is the efficiency of Cr(VI) reduction by the ferrous iron because any Cr(VI) that is not reduced will pass through the filter and contribute to total chromium. The second important measure is the efficiency of Cr(III) removal. This measurement indicates how efficiently the combination of preexisting Cr(III) and Cr(III) formed from the reduction of Cr(VI) was precipitated, coprecipitated, or coagulated by iron and removed by the filter. Third, the efficiency of the entire process can be measured by total chromium removal, which is determined by the combined efficiency of Cr(VI) reduction and Cr(III) precipitation and coprecipitation.

Table 8.2 compares the efficiency of Cr(VI) reduction, Cr(III) removal, and total chromium removal in the LADWP water as a function of pH. As shown in Table 8.2, lowering the pH improved performance in two ways. First the reduction of Cr(VI) to Cr(III) was more complete. Second Cr(III) removal appeared to become more efficient, presumably through improved coprecipitation or coagulation with iron.

It is unclear why decreasing the pH of the LADWP water to 6.5 both improved the reduction of Cr(VI) to Cr(III) and appeared to improve the coprecipitation or coagulation of Cr(III) with iron. One possible mechanism involves the potential formation of ferrous iron–silica complexes, which could inhibit both Cr(VI) reduction and Cr(III) coprecipitation or coagulation at pH values greater than 6.5. It should be remembered that in the Phase B experiments, the presence of small amounts of silica inhibited the reduction of Cr(VI) by Fe(II) (see Figure 8.3). It should also be noted that the silica concentration of water from LADWP well AT002 is about 11 mg/L as SiO_2.

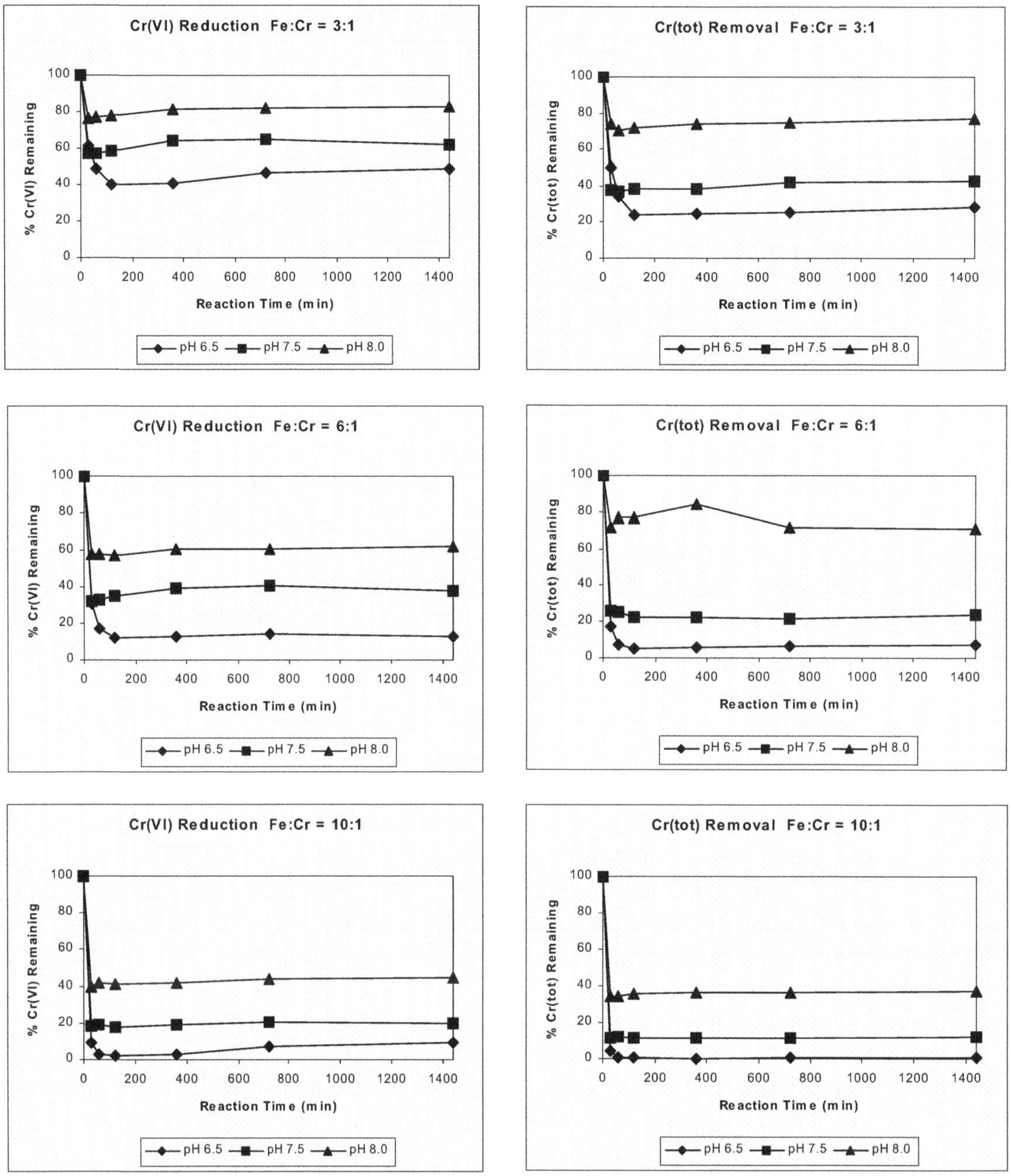

Figure 8.6 Removal of Cr(VI) and total chromium using the reduction–precipitation process in LADWP water

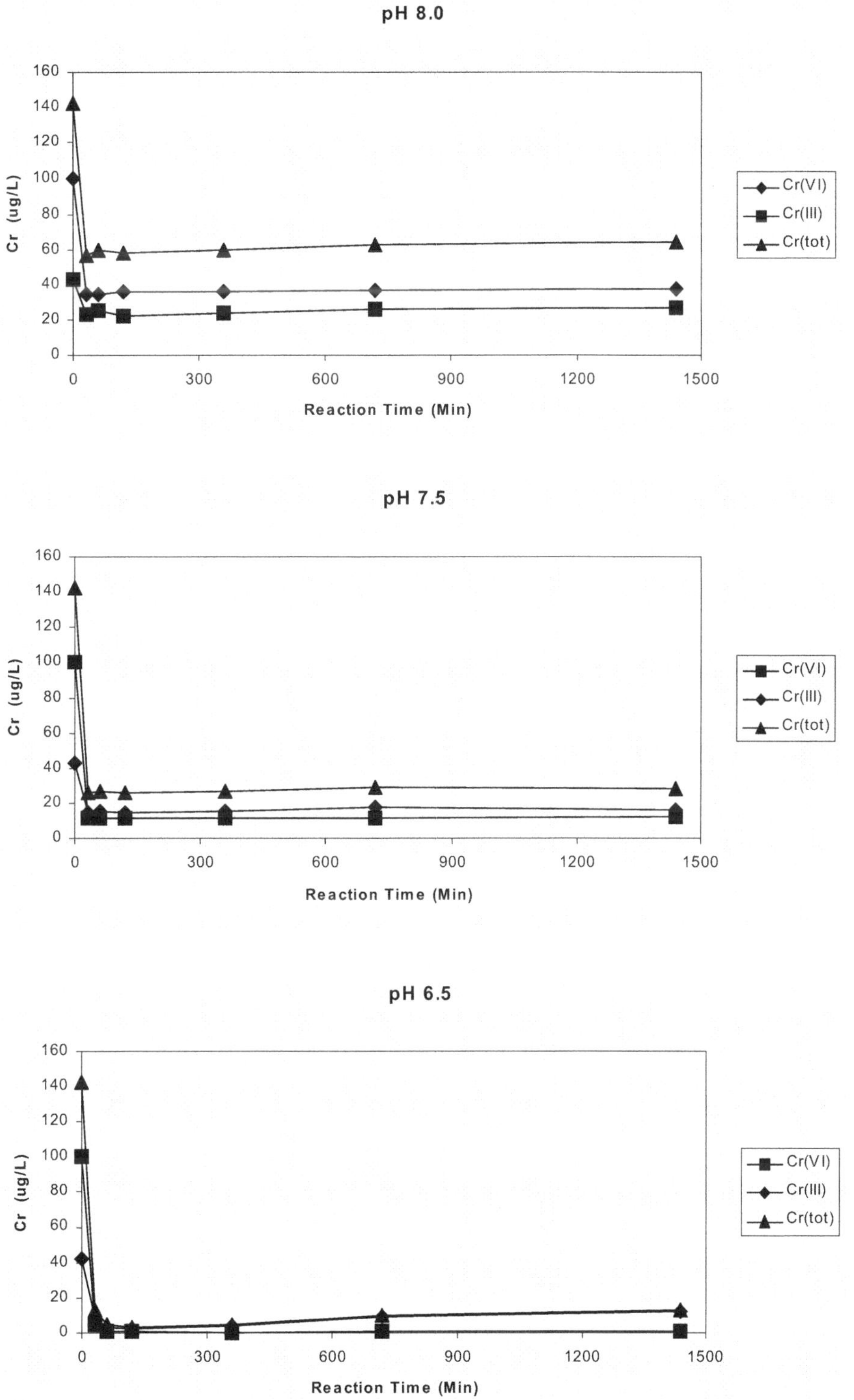

Figure 8.7 Comparison of chromium speciation in LADWP water after reduction and 0.2-µm filtration

Table 8.2
Chromium removal from LADWP water by reduction and precipitation with an Fe:Cr weight ratio of 10:1

	Initial concentration (μg/L)			Final concentration* (μg/L)			Performance efficiency (%)		
pH	Cr(VI)	Cr(III)	Total chromium	Cr(VI)	Cr(III)	Total chromium	Cr(VI) reduction	Cr(III) removal	Total chromium removal
8	100	45	145	36	25	61	64	78	58
7.5	100	45	145	12	16	28	89	88	81
6.5	100	45	145	ND	7	7	>99	95	95

* 120 min of reaction time and filtration with 0.2-μm filter

A mineral equilibrium (MINEQL) simulation using the water quality characteristics of well AT002 was performed to study the fate of ferrous iron in this system. Results of the simulation are presented in Figure 8.8. In fact, an iron silica precipitate $Fe_2Si_2O_5(OH)_4$ (Greenalite) was predicted to exist between pH 7 and 10.

Figure 8.9 describes a proposed mechanism by which silica could interfere with the Cr(VI) reduction and precipitation process. The left side of the figure illustrates the assumed treatment process: Fe(II) is oxidized to Fe(III), and Cr(VI) is reduced to Cr(III). After the oxidation–reduction reaction is completed, both iron and chromium form insoluble products that may be coprecipitated or coagulated. In the presence of silica (see right side of figure), the formation of an Fe(II) silica precipitate, Greenalite, creates a demand for ferrous iron, making Fe(II) unavailable to enter into an oxidation–reduction reaction with Cr(VI). The formation of this precipitate would also reduce the amount of ferric iron available to promote coprecipitation or coagulation. In both cases, the efficiency of Cr(VI) reduction and Cr(III) removal would be reduced. Because the formation of Greenalite predominates between pH 7.5 and 9.5, this interference with Cr(VI) reduction would not be significant at pH 6.5. If this mechanism is actually occurring, the best way to overcome it would be to add more ferrous iron to overpower the silica effect.

PHASE E—SIMULATION OF TOTAL CHROMIUM REMOVAL IN EXPERIMENTS WITH A NATURAL WATER SOURCE SUPPLIED BY GLENDALE

The performance of Cr(VI) reduction and precipitation using Fe(II) in a chromium-contaminated natural water source supplied by the City of Glendale was evaluated. As in the experiments with LADWP water, the process's performance was evaluated as a function of reductant dose and pH. The Fe(II) reductant was added at an Fe:Cr weight ratio of 10:1 and 100:1, and the experiments were conducted at pH 8.0, 7.5 and 6.5. After the reduction reaction was initiated, aliquots were taken after 120 min and fractionated by passing the aliquot through 0.45-μm, 0.2-μm, 0.02-μm, 100,000-dalton, 10,000-dalton, and 1000-dalton membrane filters. The effect of a coagulant aid, added to the water prior to filtration, was also evaluated. Polymer dosages of 0.5 mg/L and 1.0 mg/L were evaluated.

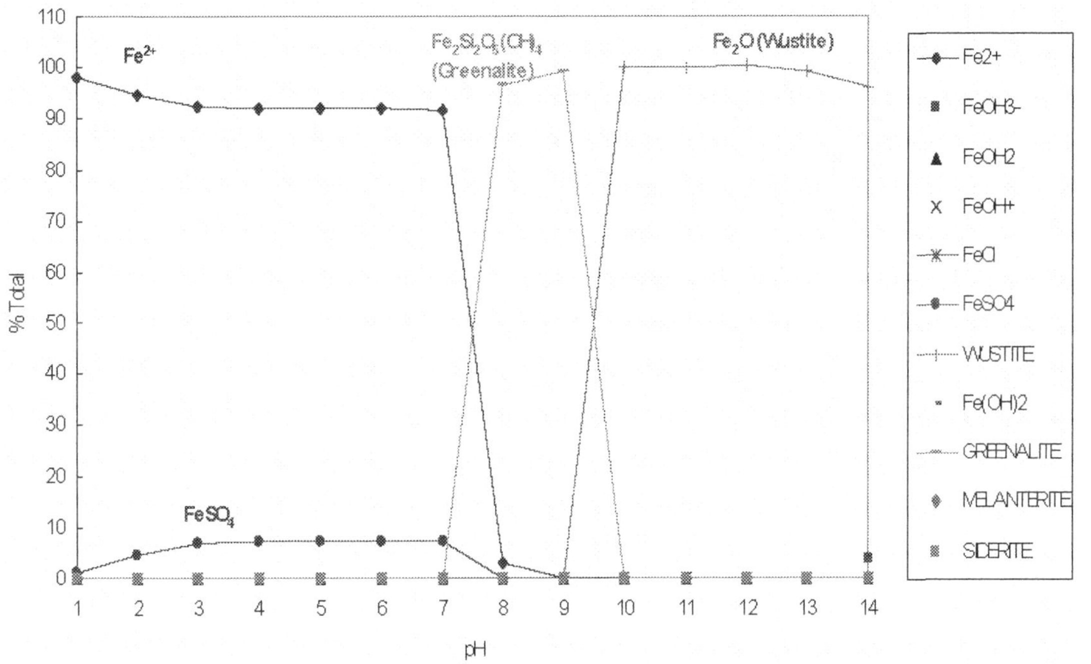

Figure 8.8 MINEQL simulation of Fe(II)-complexed species as a function of pH in water from LADWP well AT002

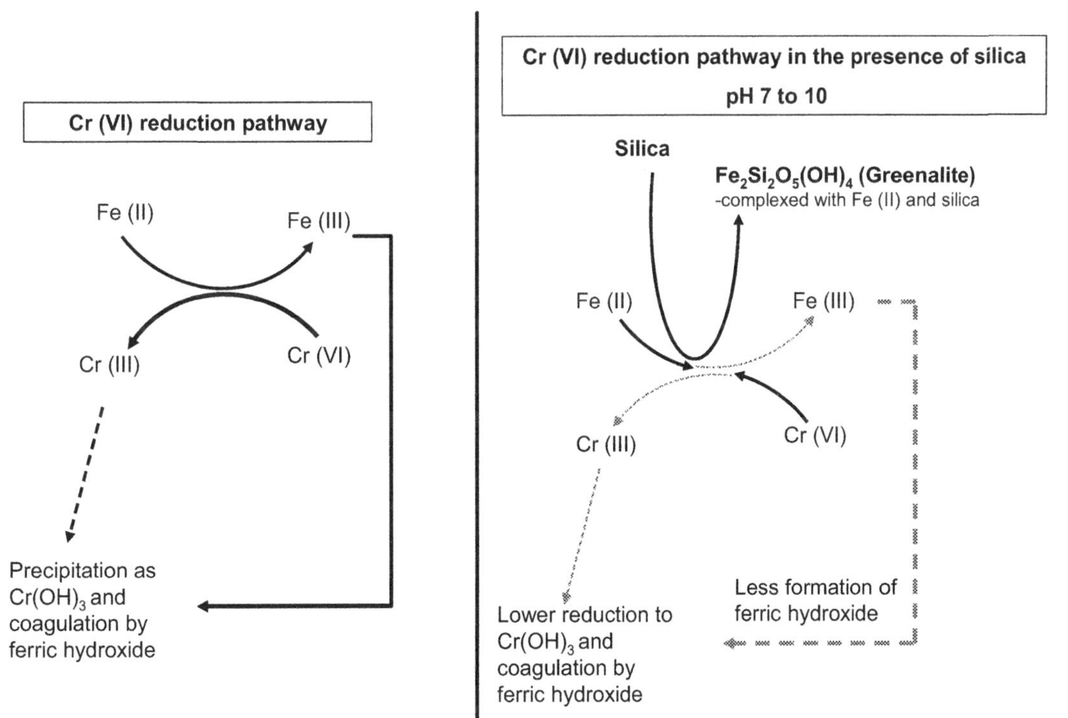

Figure 8.9 Possible inhibition pathway for Cr(VI) reduction by Fe(II) and coprecipitation or coagulation of Cr(II) in the presence of silica

The water used for the Phase E experiments was taken from the effluent of the Glendale water treatment plant prior to chlorination. The water is a blend of supplies from eight wells contaminated with trace levels of Cr(VI). Water quality characteristics of the blended water are given in Table 8.3. As received, the water contained approximately 5 μg/L total chromium and 5 μg/L Cr(VI), indicating that the chromium in this sample was totally speciated as Cr(VI). Before the experiments were run, the water was spiked to an initial Cr(VI) concentration of 100 μg/L. All experiments were performed open to the atmosphere and at room temperature.

Results of the reduction–precipitation process in Glendale water are presented in Figure 8.10. As with LADWP water, the lowest total chromium concentration was observed when Cr(VI) reduction occurred at pH 6.5. Very little difference in total chromium removal was attributable to filter pore size at a given pH, indicating that the portion of total chromium that is not removed by filtration is less than 1000 daltons and is truly dissolved. Increasing the Fe:Cr weight ratio by a factor of 10—from 10:1 to 100:1—improved total chromium removal by all filters under all pH conditions, but the improvement was greater at pH 8.0 and 7.5 than at pH 6.5. At pH 8.0 and 7.5, total chromium removal by all filters was improved by a factor of approximately three. At pH 6.5, total chromium removal by all filters improved by a factor of approximately two. As a point of comparison, decreasing the pH from 7.5 to 6.5 provided roughly the same level of improvement in total chromium removal as a 10-fold increase in Fe(II) dose. However, even after treatment at an Fe:Cr weight ratio of 100:1 and a lowered pH of 6.5, approximately 3 μg/L of total chromium passed through a 1000-dalton filter. In other words, approximately 3% of the initial chromium concentration in the water appeared to remain in solution.

An experiment was conducted to determine if the addition of a polymer could improve total chromium removal by potentially improving the aggregation of precipitated or coprecipitated Cr(III) into colloids or particles large enough to be removed by the membrane filter. A high-molecular-weight high-anionic-charge polymer, Ciba Magnafloc LT26, was used. The experiment was performed at pH 6.5 and at an Fe:Cr weight ratio of 100:1. Two polymer dosages, 0.5 mg/L and 1.0 mg/L, were evaluated. Results of the experiment are plotted in Figure 8.11. As shown in the figure, addition of the polymer did not improve total chromium removal under the conditions of this experiment.

Table 8.3
Glendale treatment plant water quality (fall 2003)

Parameter	Unit	Value	Parameter	Unit	Value
pH	-	7.3	Mg^{2+}	mg/L	29.2
TDS	mg/L	375	Na^{+}	mg/L	40.6
Cr(VI)	μg/L	7	K^{+}	mg/L	2.4
Cr(tot)	μg/L	7	Cl^{-}	mg/L	64.3
SiO_2	mg/L	17.1	SO_4^{2-}	mg/L	105.7
Ca^{2+}	mg/L	87.9	NO^{3-}	mg/L	6

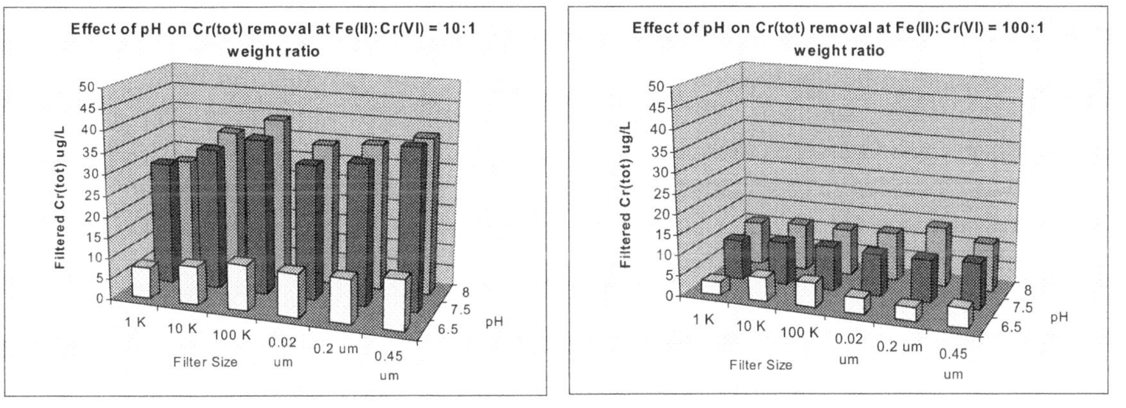

Figure 8.10 Comparison of filtered total chromium concentrations after reduction with Fe(II) in Glendale water spiked to 100 µg/L Cr(VI)

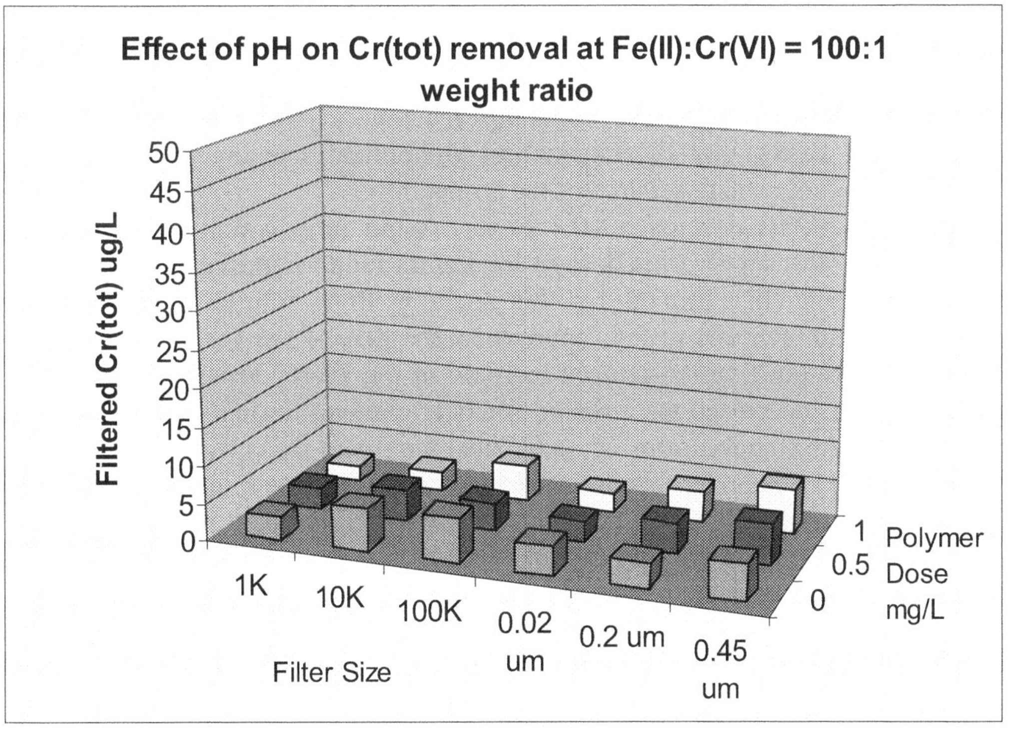

Figure 8.11 Influence of polymer addition on filtered total chromium concentration (pH = 6.5)

SUMMARY

Treatment of Cr(VI) by reduction and precipitation was studied through a series of bench-scale experiments. Initial experiments compared the performance of Fe(II) and Fe(0) as reductants in synthetic water supplies. Under the conditions tested in the synthetic water, zero-valent iron was found to be ineffective at reducing Cr(VI). Cr(VI) reduction by Fe(0) was not tested in natural water sources. All subsequent experiments simulating the reduction–precipitation process used Fe(II) as the reductant.

Fe(II) was an effective reductant for Cr(VI) in both synthetic and natural water supplies. In the synthetic water at a pH of 7.5, a Fe(II) dose of at least a 6:1 Fe:Cr weight ratio (two times the stoichiometric dose) was required to effectively reduce Cr(VI) to Cr(III). Cr(VI) reduction in the synthetic water supply was fairly rapid, with the bulk of the reduction completed within 10 min. In two chromium-contaminated natural water sources at pH 6.5, an Fe(II) dose of at least a 10:1 Fe:Cr weight ratio (three times the stoichiometric dose) was needed to reduce Cr(VI) to Cr(III). Cr(VI) reduction in the natural water supply took approximately 60 min to complete.

The influence of pH on Cr(VI) reduction by Fe(II) was dissimilar in the synthetic and natural water supplies. In the synthetic water, increasing the pH from 5.5 to 7.5 improved the rate and extent of the reduction reaction. In contrast, reducing the pH from 8.0 to 6.5 improved the extent of the chromium reduction reaction in both natural water supplies.

The presence of silica inhibited Cr(VI) reduction by Fe(II) in the synthetic water at a pH of 7.5. This result is surprising because silica is not directly involved in the exchange of electrons that occurs during the oxidation–reduction reaction between Cr(VI) and Fe(II). It is speculated that silica indirectly interferes with the reduction of Cr(VI) through the formation of a ferrous silica precipitate. The formation of this compound would consume ferrous ions, making them unavailable to participate in the Cr(VI) reduction reaction. MINEQL simulations indicated that the iron silica solid $Fe_2Si_2O_5(OH)_4$ (Greenalite) predominates in the pH range of 7.5 to 9.5. Formation of the solid may be the source of the competition.

Although an Fe(II) dose of a 10:1 Fe:Cr weight ratio completely reduced Cr(VI) to Cr(III) in a natural water supply, this dosage apparently failed to provide a sufficient amount of iron to remove the Cr(III) formed by the reaction from solution via coprecipitation or coagulation. At the ambient pH of 7.5, approximately 40% of the Cr(III) formed by reduction passed through a 0.45-µm filter. A tenfold increase of the Fe(II) dosage, from an Fe:Cr weight ratio of 10:1 to 100:1, decreased the amount of Cr(III) passing through a 0.45-µm filter to 10%. Lowering the pH of the solution from 7.5 to 6.5 also improved the performance of the reduction–precipitation process. Even under this optimized condition, however, approximately 3–5% of the reduced chromium remained in solution. This potentially indicates a performance threshold (at the 2- to 5-µg/L range) that the process cannot exceed.

In conclusion, the reduction–precipitation process using ferrous iron as a reductant appears to be a feasible method for treating chromium in drinking water sources to the low microgram-per-liter range.

CHAPTER 9
ASSESSMENT OF CHROMIUM TREATMENT TECHNOLOGIES

SUMMARY OF THE PERFORMANCE OF CHROMIUM TREATMENT TECHNOLOGIES

Chromium's complex chemistry allows for several alternative approaches to treating Cr(VI) in drinking water. Figure 9.1 presents an array of treatment alternatives that may be capable of controlling Cr(VI) at very low concentrations. As indicated in Figure 9.1, most of the technologies were evaluated at bench scale as part of this project.

Two basic approaches are available for treating chromium—treating either the oxidized Cr(VI) species or the reduced Cr(III) species. Of the technologies considered by this project, only RO/NF membranes were effective at simultaneously treating both chromium species. Although this work focused on the treatment of Cr(VI), a more complete picture is required. The potential for Cr(III) oxidation in distribution systems necessitates removing total chromium from the treatment stream, not merely changing the oxidation state to a form that poses no health hazard [Cr(III)]. At a minimum, chromium must be treated to a level at which the complete conversion of chromium to the Cr(VI) state in a distribution system would not exceed the treatment objective.

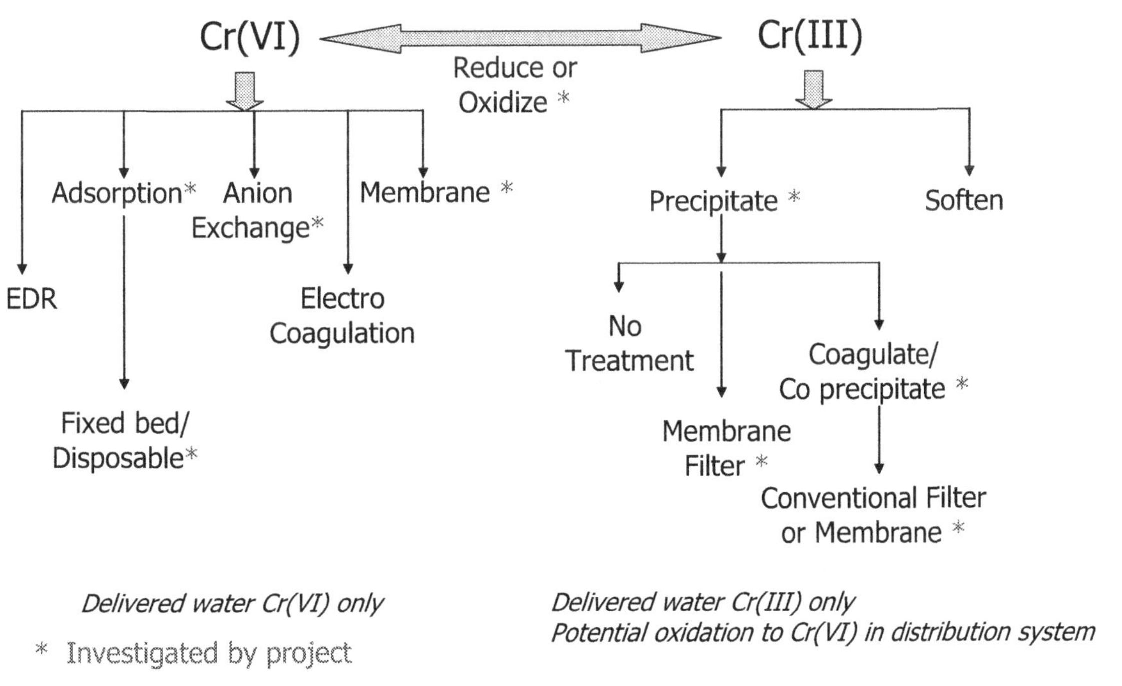

Figure 9.1 Alternative methods for controlling chromium

Mechanistically, treatment technologies that can effectively remove Cr(VI) take advantage of its anionic properties. Technologies that remove Cr(III) at moderate pH values take advantage of its low solubility. However, one conclusion of this bench-scale work is that it appears that a fraction of Cr(III) may remain soluble and may be difficult to remove. Because it is very small and uncharged, Cr(III) cannot easily be removed from water. At drinking water pHs, an effective way to achieve very low chromium concentrations may be to treat chromium in its anionic Cr(VI) form.

Table 9.1 provides an overview of anticipated levels of performance by the alternative chromium treatment technologies considered in this project. Because performance levels were drawn from bench-scale tests, they represent an estimate of performance at a larger scale. Nonetheless, Table 9.1 provides guidance for selecting Cr(VI) treatment technologies for further evaluation.

Bench-scale tests indicated that two technologies appear to be able to control chromium to very low (<2 μg/L) concentrations: conventional anion exchange and RO. Both technologies are used somewhat in the drinking water field and have the advantage of being able to treat multiple contaminants. Unfortunately, although both technologies are extremely effective at treating chromium, they create residual streams that can be difficult to dispose of.

Four treatment processes investigated by the project were able to achieve low (2–10 μg/L) effluent chromium concentrations. These include reduction followed by coprecipitation–coagulation, the continuously mixed anion exchange process (MIEX®), NF, and a granular media treatment process, sulfur-modified iron (SMI), which appears to operate by means of surface reduction and precipitation. Reduction followed by coprecipitation–coagulation is widely used as an industrial wastewater treatment process and is easily applicable to drinking water treatment. NF offers an effective alternative to RO, but its ability to reject Cr(VI) is much more sensitive to water quality conditions than RO. SMI appears to have potential as a Cr(VI) treatment process and is the only process deemed effective that operates as a disposable media. However, though apparently effective at chromium control, SMI is an immature product that requires further development and commercialization. SMI is a product worthy of additional investigation.

Several treatment processes were deemed ineffective at controlling chromium. Surprisingly, disposable iron-based sorption media, developed for arsenic treatment, were ineffective at controlling Cr(VI). In contrast to RO and NF, UF was also ineffective at Cr(VI) rejection in natural water supplies. Given the large pore size of UF membranes, preventing the passage of Cr(VI) is not possible. This work has shown that rejection of Cr(VI) by electrostatic repulsion by the NF membrane surface is not an effective mechanism in the natural water supplies studied. Last, the potential for reoxidation in the distribution system precludes treatment processes that only reduce Cr(VI) to Cr(III) and do nothing to remove Cr(III) from the treated water.

Two additional potentially effective treatment processes were identified but not evaluated by the project: electrodialysis–electrodialysis reversal (ED–EDR) and electrocoagulation. Based on the performance of ED–EDR in treating other oxyanions, it was anticipated that the performance of this process would be similar to that of RO. However, because this technology is infrequently used in drinking water treatment (and because no vendor was willing to provide a unit), no evaluation was conducted. Softening was not evaluated based on results reported in the literature and indicating poor removal of chromium by softening processes (Sorg 1979).

Table 9.1
Performance of Cr(VI) treatment technologies, based on bench-scale testing

Capable of controlling Cr(VI) to very low effluent concentrations (<2 μg/L)	
Conventional Anion Exchange	Highly effective, mature technology.
Reverse Osmosis	Highly effective, mature technology.
Capable of controlling Cr(VI) to low effluent concentrations (2–10 μg/L)	
Reduction coprecipitation–coagulation	Performance appears to be limited by Cr(III) solubility
Continuously mixed anion exchange	Unable to maintain very low effluent concentrations under conditions tested; additional study warranted
Nanofiltration	Performance related to membrane surface charge and can be limited by water quality
Surface reduction–precipitation	Immature technology (sulfur-modified iron) but shows promise and warrants additional study
Not effective	
Adsorption media	Media developed for arsenic treatment were ineffective at treating chromium.
Ultrafiltration	Pore size too large to reject Cr(VI)
Reduction or reduction followed by filtration	Potential for chromium reoxidation in distribution system limits applicability
Not tested	
Electrodialysis–electrodialysis reversal	Uncommon for drinking water application but performance likely to be similar to that of RO
Electrocoagulation	Unable to quantify performance (vendor withdrew from evaluation)
Softening	Literature review indicated relatively poor performance

RECOMMENDATIONS FOR PILOT-TESTING

Chapters 5–8 provide details of the results of bench-scale testing of several treatment technologies and specific products provided by a number of vendors. This screening successfully identified several technologies that are most promising for application by drinking water utilities. The treatment processes discussed in this section are recommended for pilot-scale testing in the next phase of this project.

- **Anion exchange** (both as fixed-bed and dispersed-contactor applications). The effect of regeneration on anion exchange performance and the feasibility of operating anion exchange on a "throw-away" basis are issues that can be addressed in the next phase of study.

- **Sulfur-modified iron sorption media**. The mechanism of Cr(VI) control by this technology needs to be better understood—Cr(VI) sorption or Cr(VI) reduction to Cr(III) with removal by the iron precipitates. Post-treatment requirements for iron control also need to be investigated.
- **Membrane treatment by nanofiltration and reverse osmosis**. Rejection of Cr(VI) by membrane technologies is excellent, but the operational conditions (productivity, fouling, membrane life) that affect the feasibility of this technology may require further evaluation. Membrane treatment will not be carried forward to pilot-scale studies because of the large loss of water associated with this technology.
- **Coagulation and precipitation of reduced Cr(III)**. Reduction and removal of chromium as Cr(III) can achieve reasonably low concentrations of chromium in treated water. In combination with a post-treatment and disinfectant regime, this strategy could prove extremely effective for controlling Cr(VI) at very low concentrations. Further study of mixing conditions and process kinetics is required to optimize this process for reducing Cr(VI) to very low concentrations.

APPENDIX A
CO-OCCURRING CONTAMINANTS IN WATER SUPPLIES WITH HIGH AND LOW CHROMIUM CONCENTRATIONS

Low-Cr(tot) water = <10 μg/L, High-Cr(tot) water = >10 μg/L

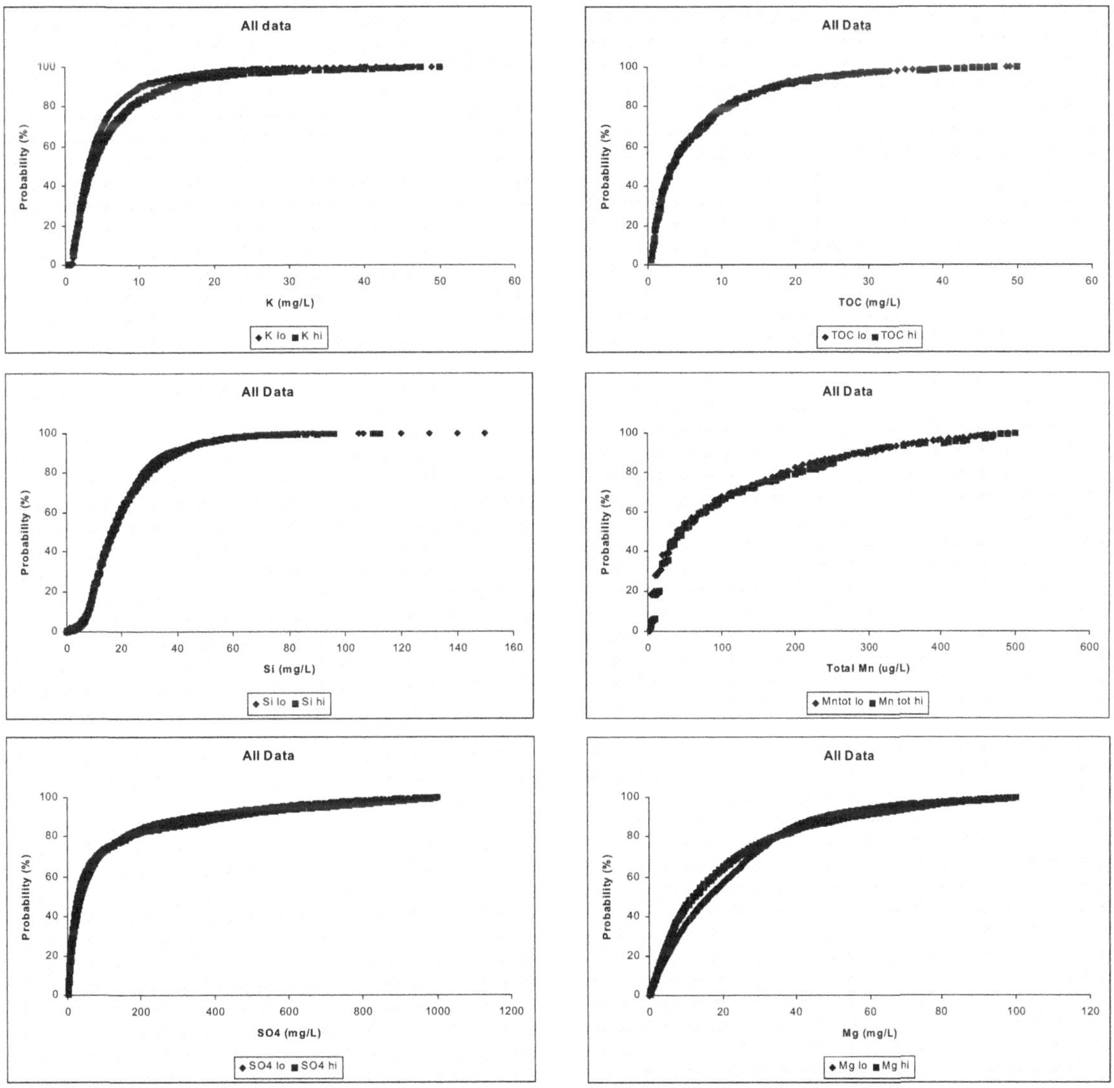

Figure A.1 Comparison of co-occurring contaminants in high- and low-chromium water supplies

(continued)

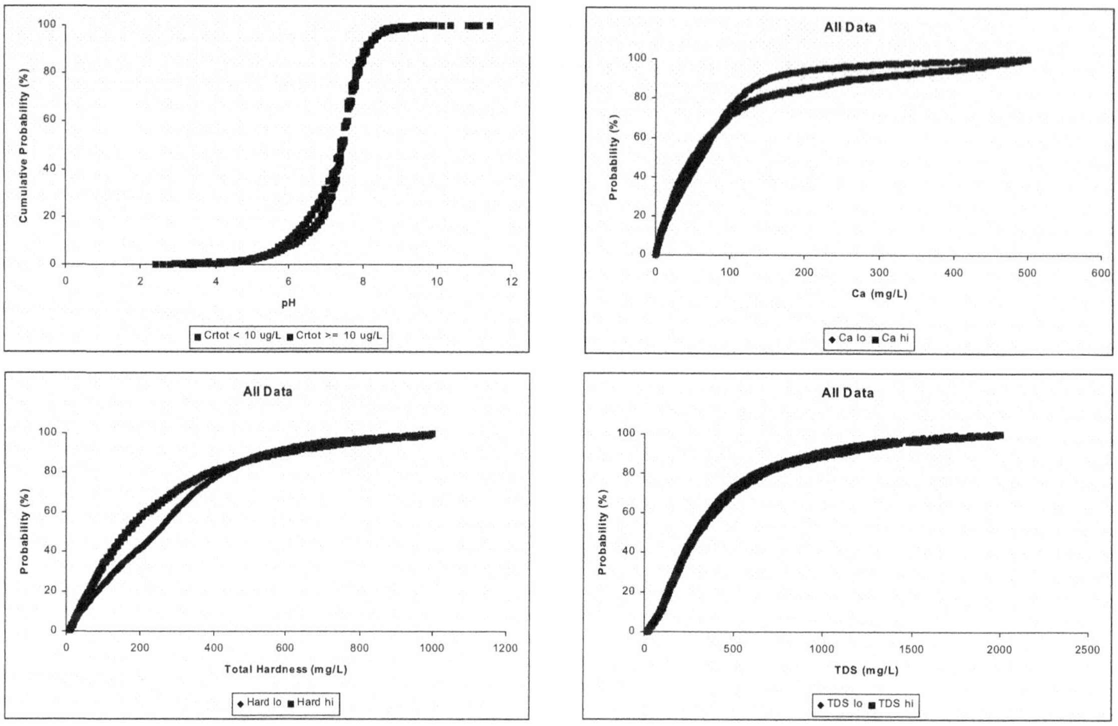

Figure A.1 (Continued)

APPENDIX B
CHROMIUM REDUCTION AND OXIDATION CHEMISTRY—SUPPLEMENTAL RESULTS

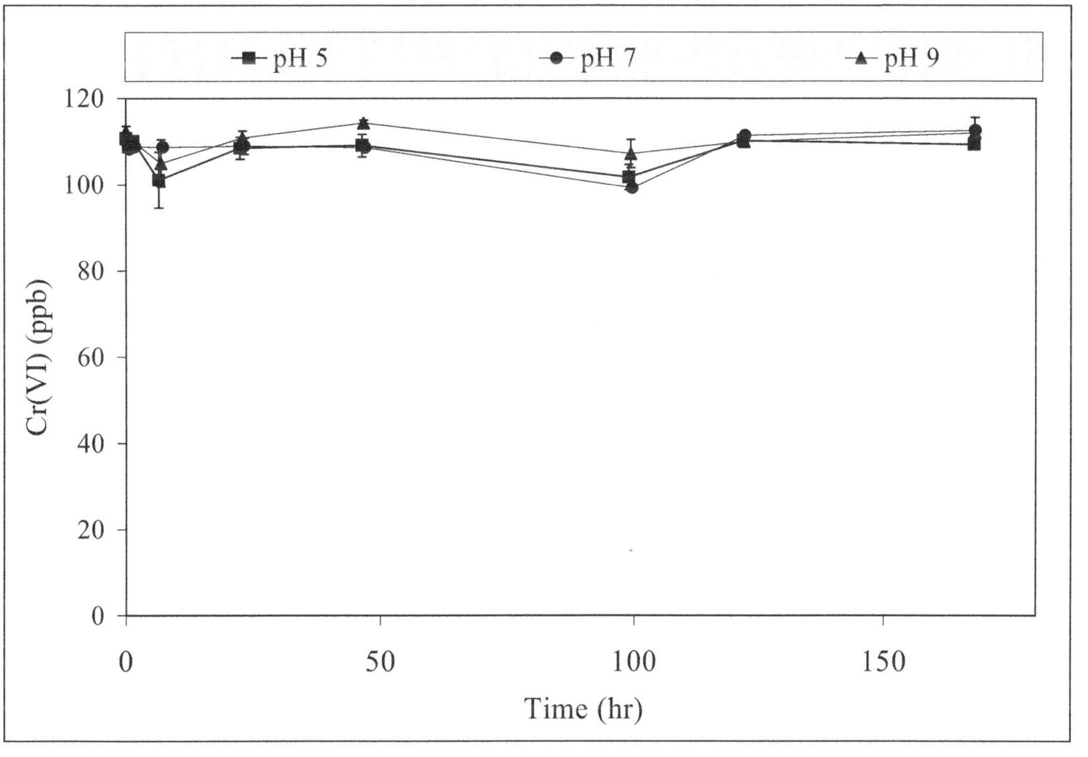

Figure B.1 Control experiment for deionized water spiked with 100 µg/L Cr(VI) (total chromium was not measured)

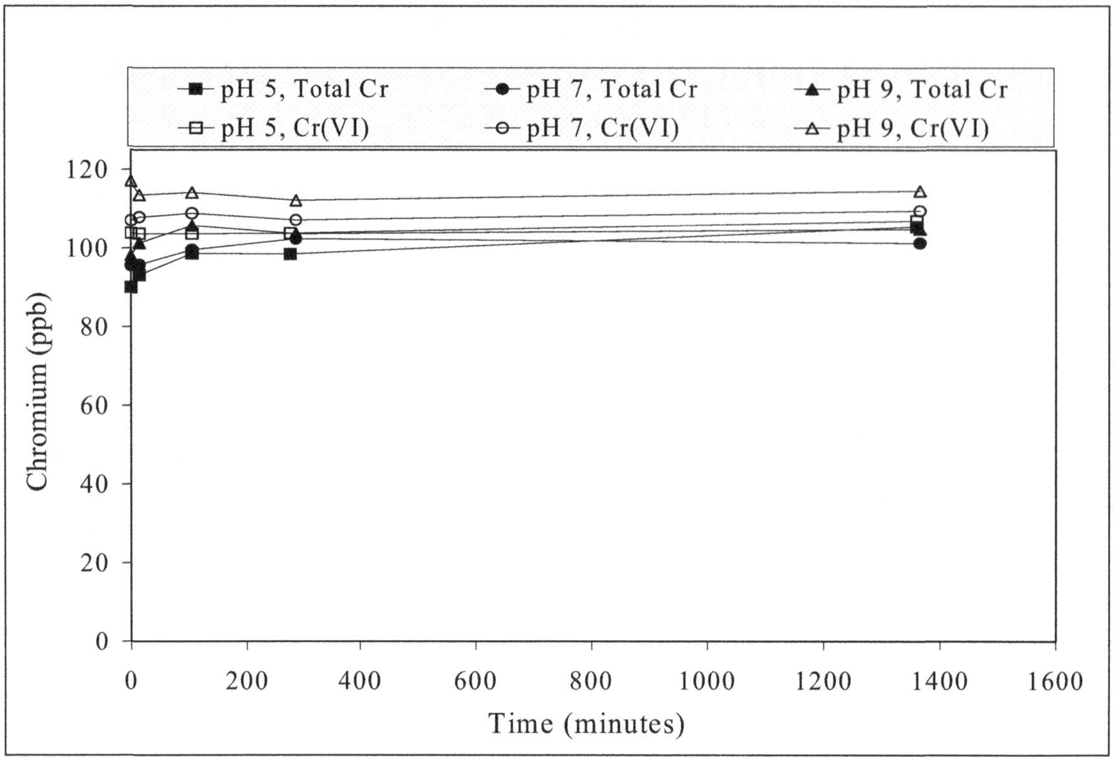

Figure B.2 Control experiment for Glendale water spiked with 100 µg/L Cr(VI) (concentrations of filtered and unfiltered samples were the same)

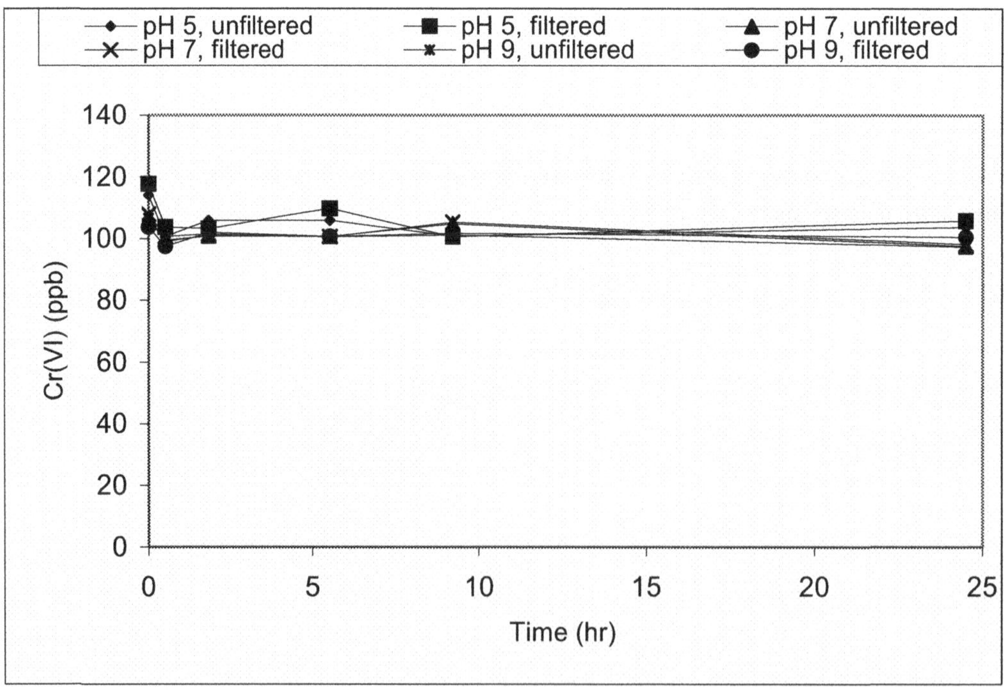

Figure B.3 Control experiment for modified synthetic water spiked with 100 µg/L Cr(VI)—total chromium and Cr(VI) concentrations were the same (data not shown)

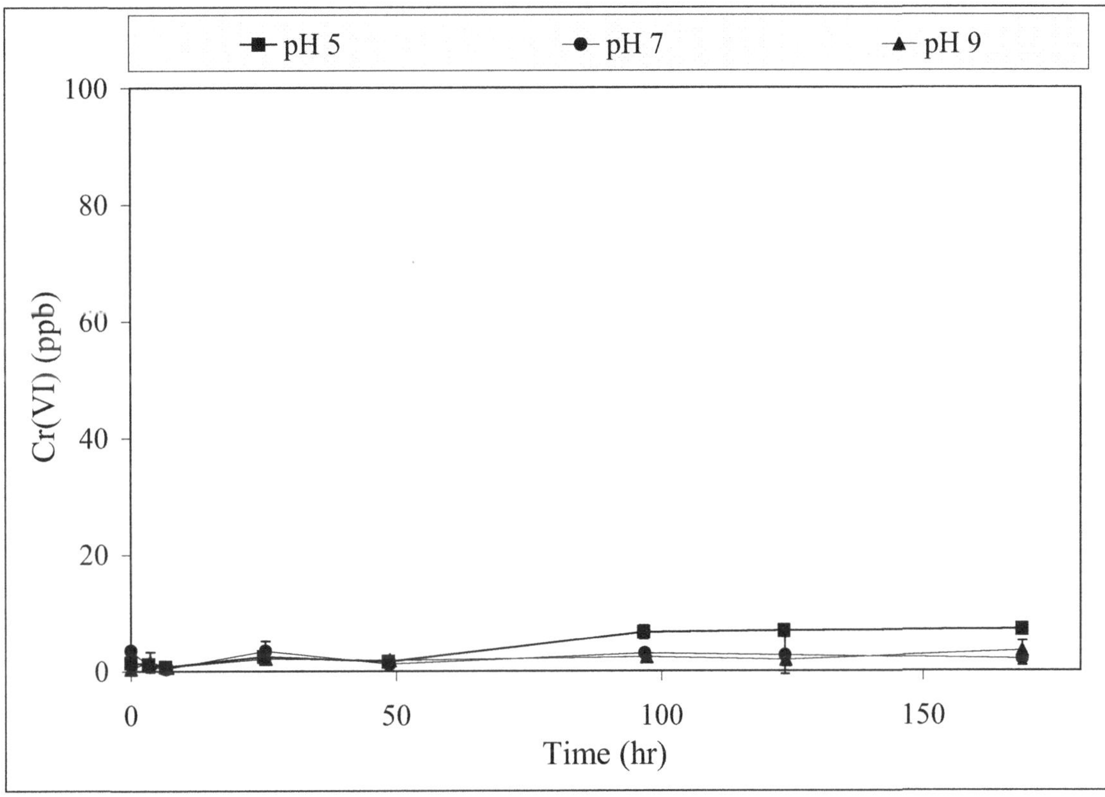

Figure B.4 Control experiment for deionized water spiked with 100 µg/L Cr(III)

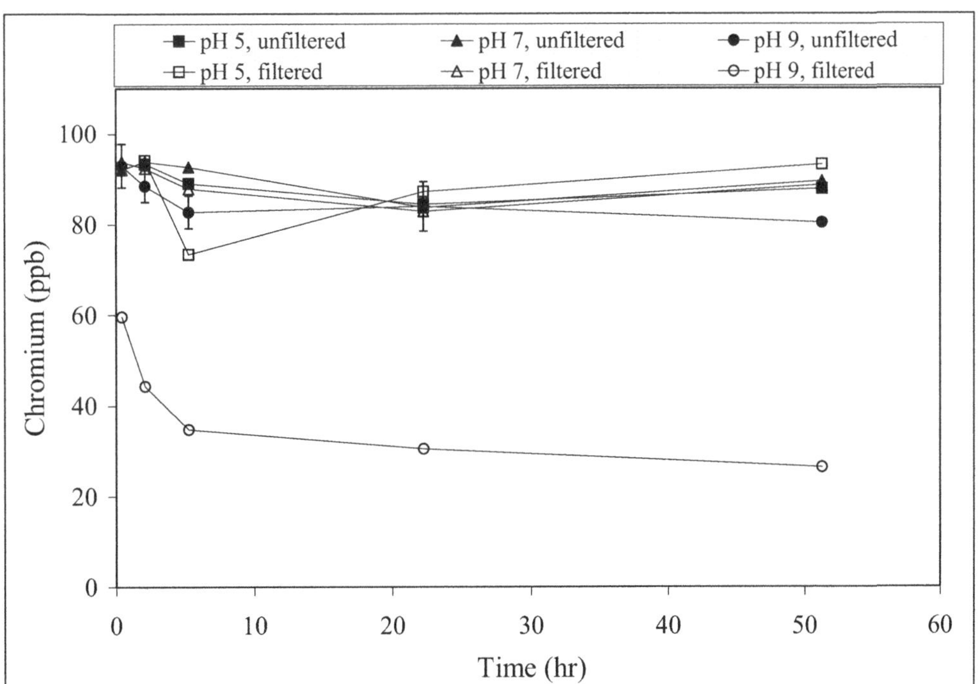

Figure B.5 Control experiment for reducing water spiked with 100 µg/L Cr(III) (no Cr(VI) was detected in any sample)

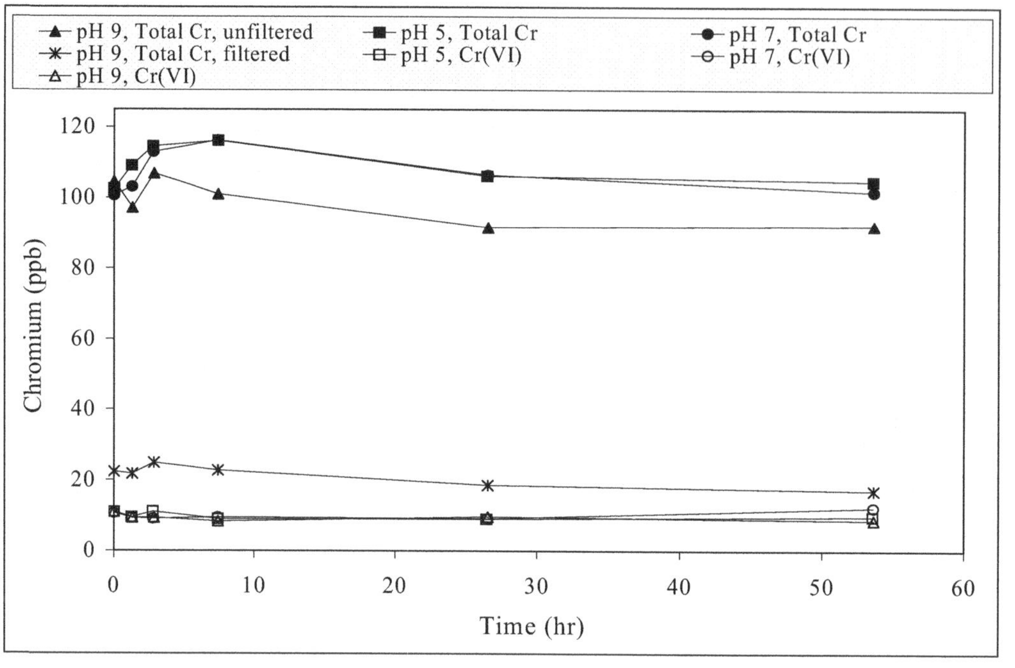

Figure B.6 Control experiment for Glendale water spiked with 100 μg/L Cr(III)

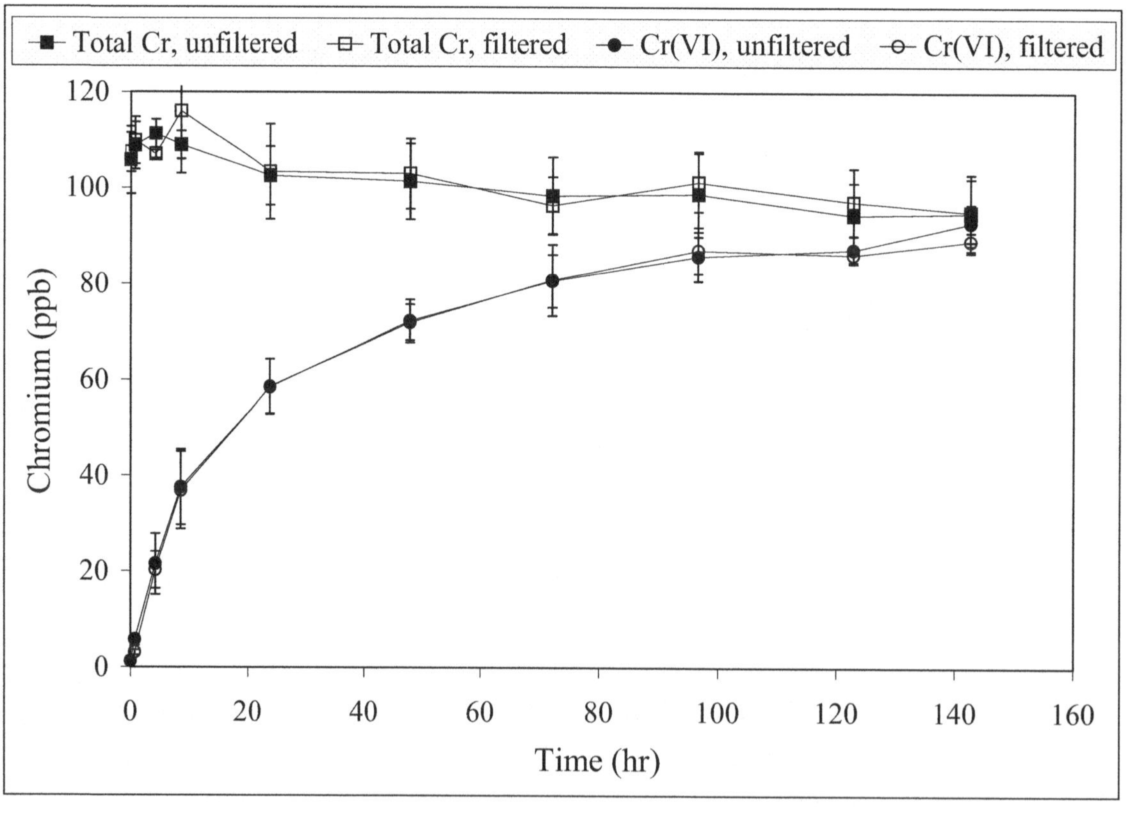

Figure B.7 Oxidation of 100 μg/L Cr(III) by 1 mg/L Cl_2 in synthetic water at pH 5

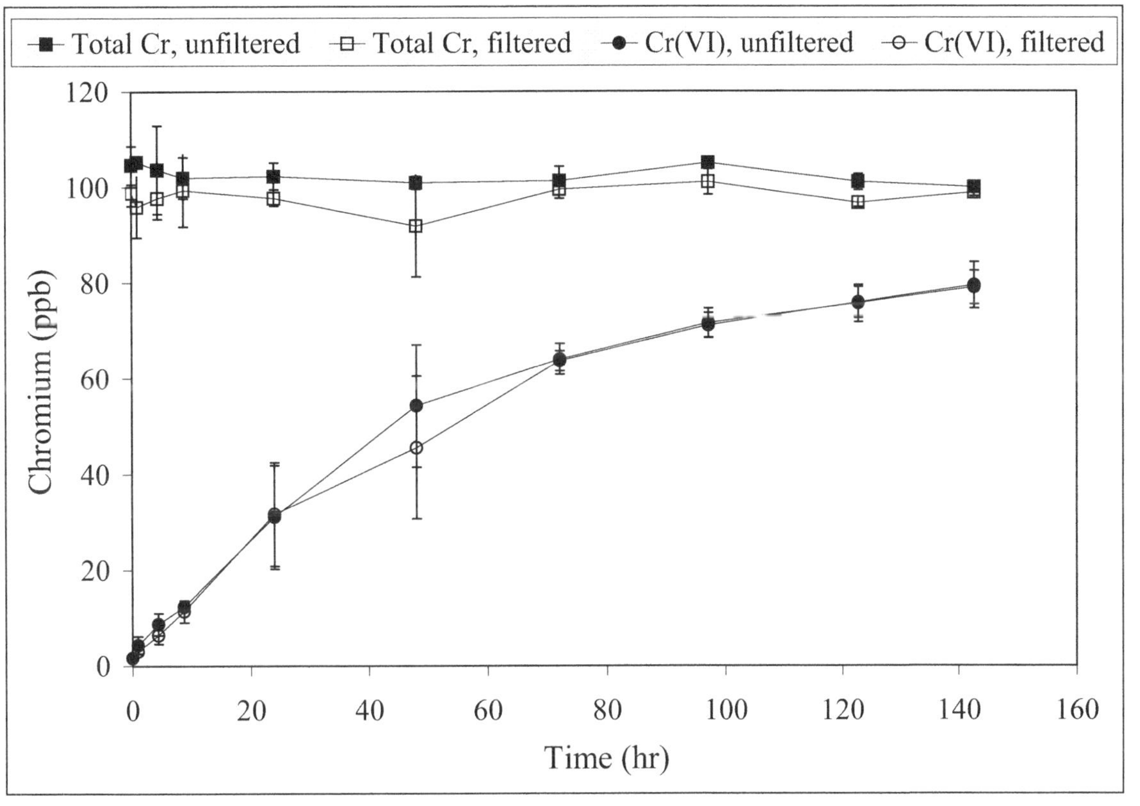

Figure B.8 Oxidation of 100 µg/L Cr(III) by 1 mg/L Cl_2 in synthetic water at pH 7

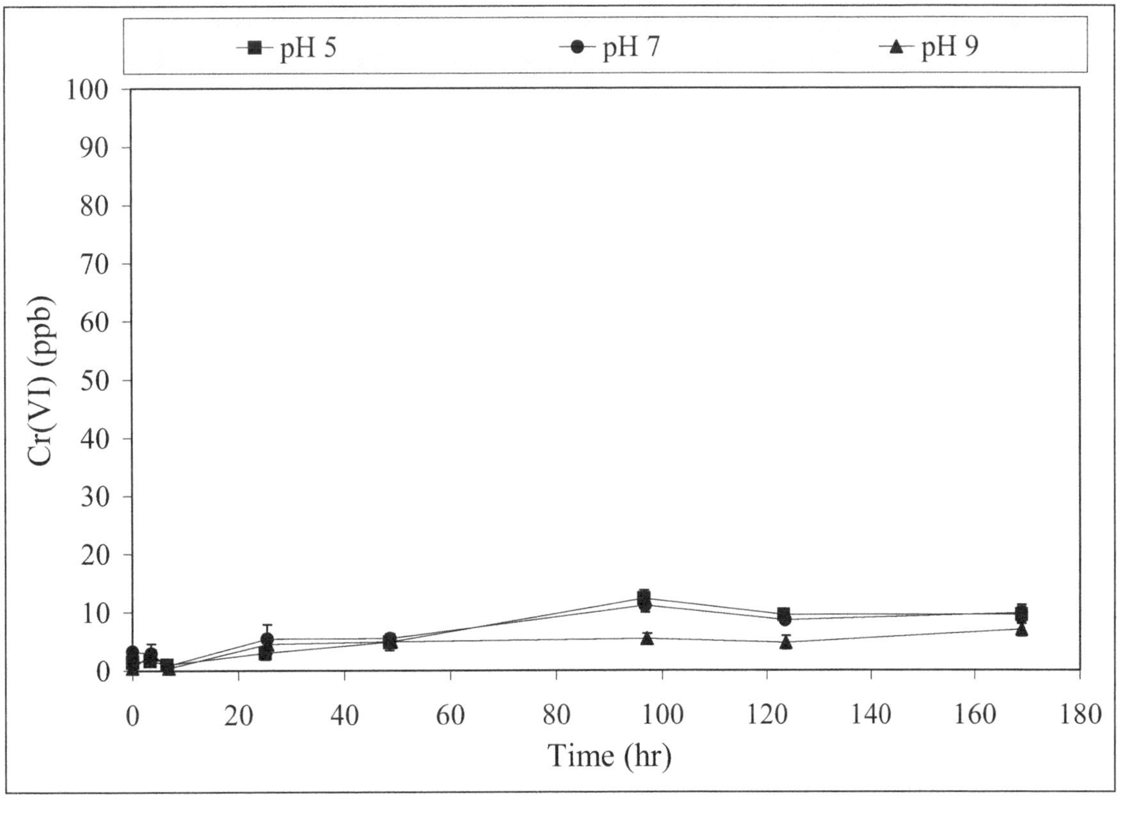

Figure B.9 Oxidation of 100 µg/L Cr(III) by 1 mg/L chloramine in deionized water

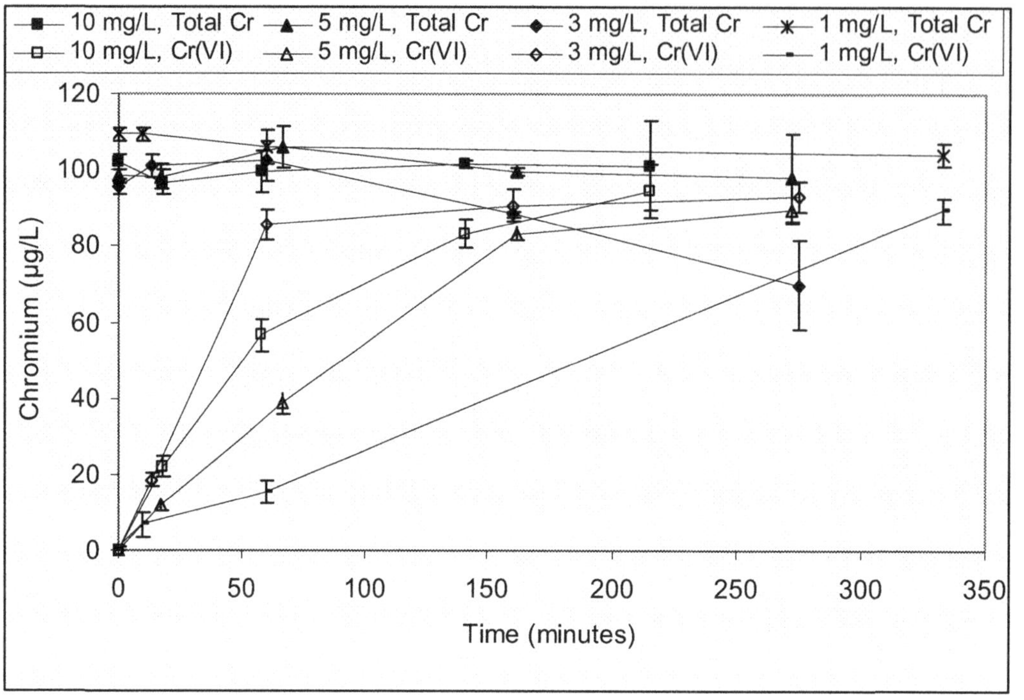

Figure B.10 Oxidation of 100 µg/L Cr(III) by potassium permanganate in synthetic water at pH 5 (all samples are unfiltered)

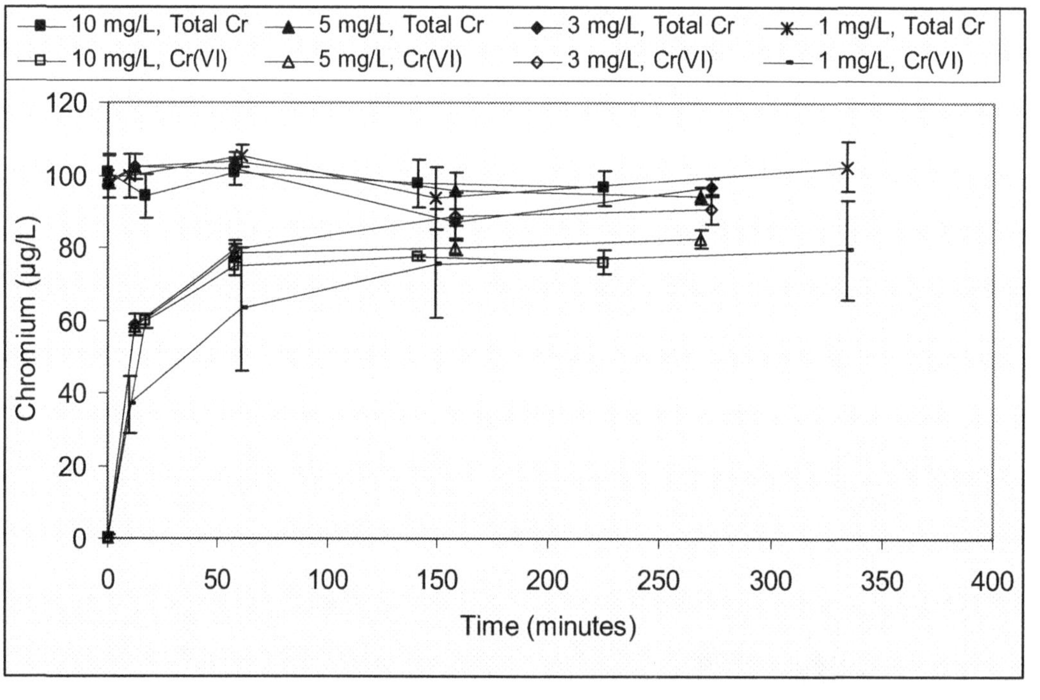

Figure B.11 Oxidation of 100 µg/L Cr(III) by potassium permanganate in synthetic water at pH 7 (all samples are unfiltered)

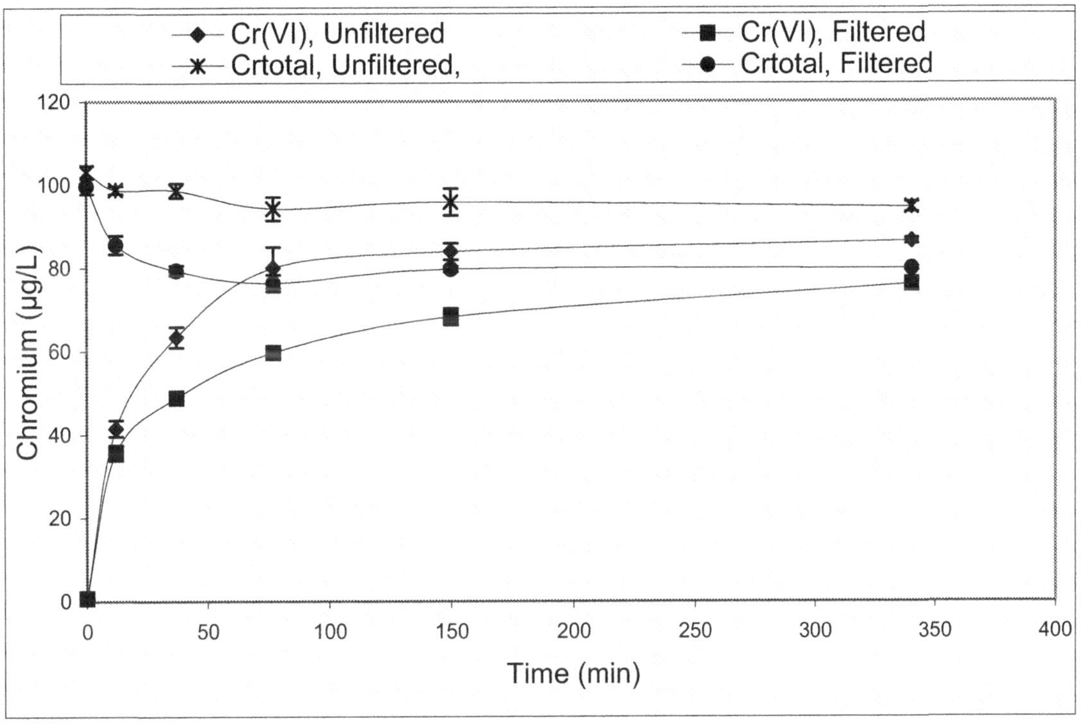

Figure B.12 Oxidation of 100 µg/L Cr(III) by 3 mg/L potassium permanganate in reducing water at pH 5

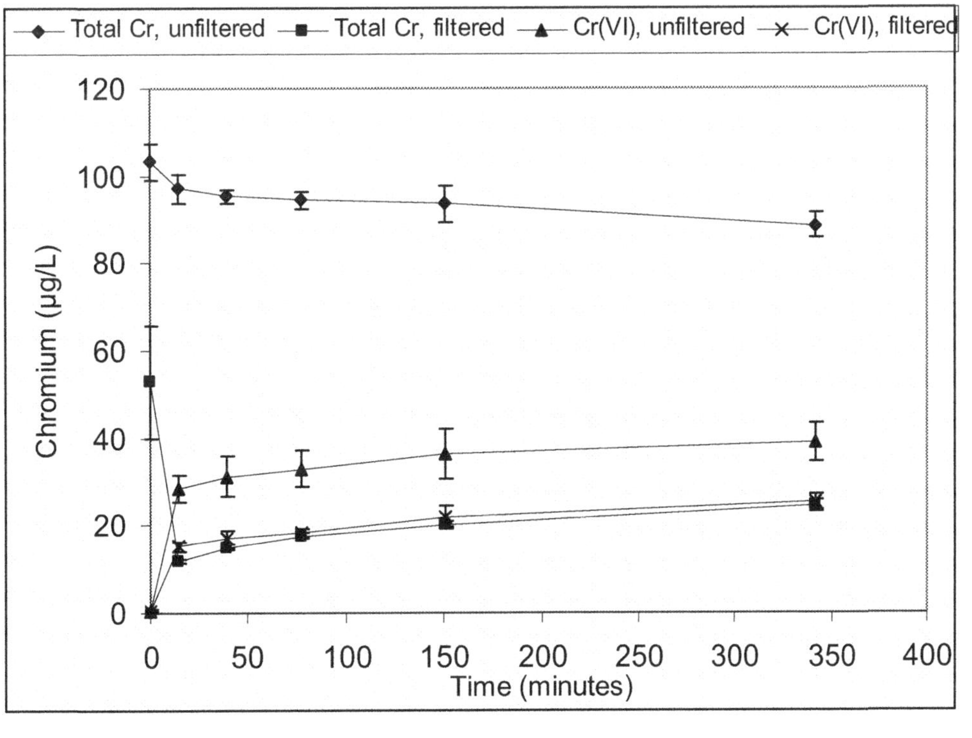

Figure B.13 Oxidation of 100 µg/L Cr(III) by 3 mg/L potassium permanganate in reducing water at pH 9

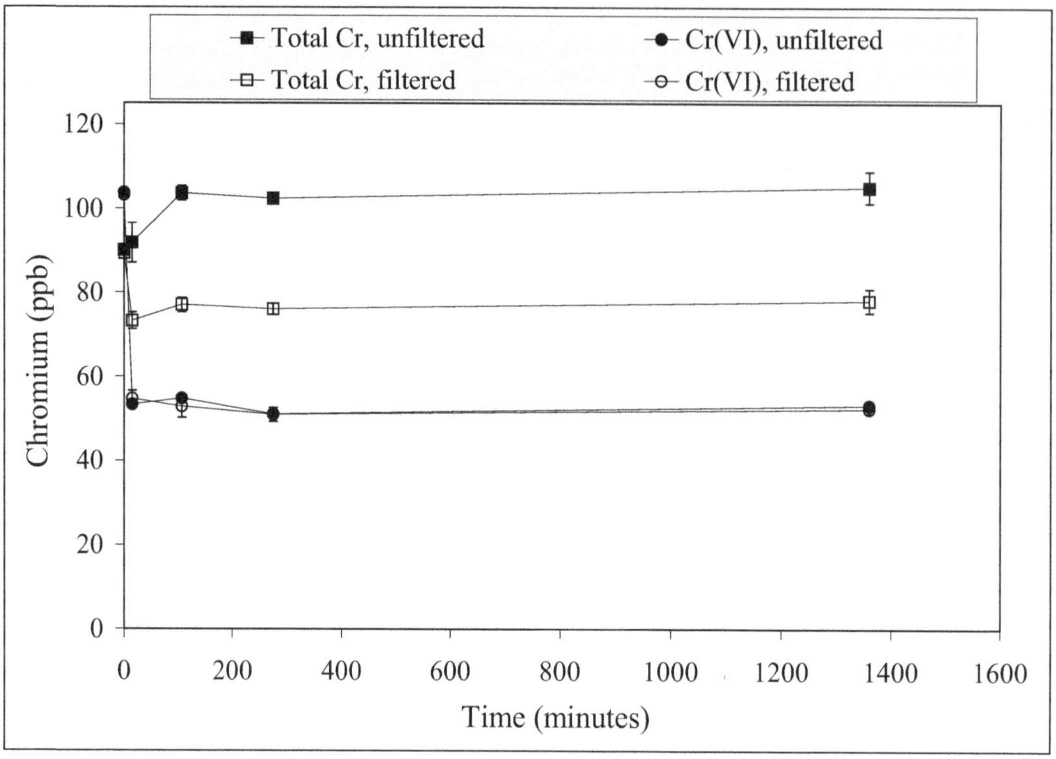

Figure B.14 Reduction of 100 µg/L Cr(VI) by 1.3 mg/L stannous chloride in Glendale water at pH 5

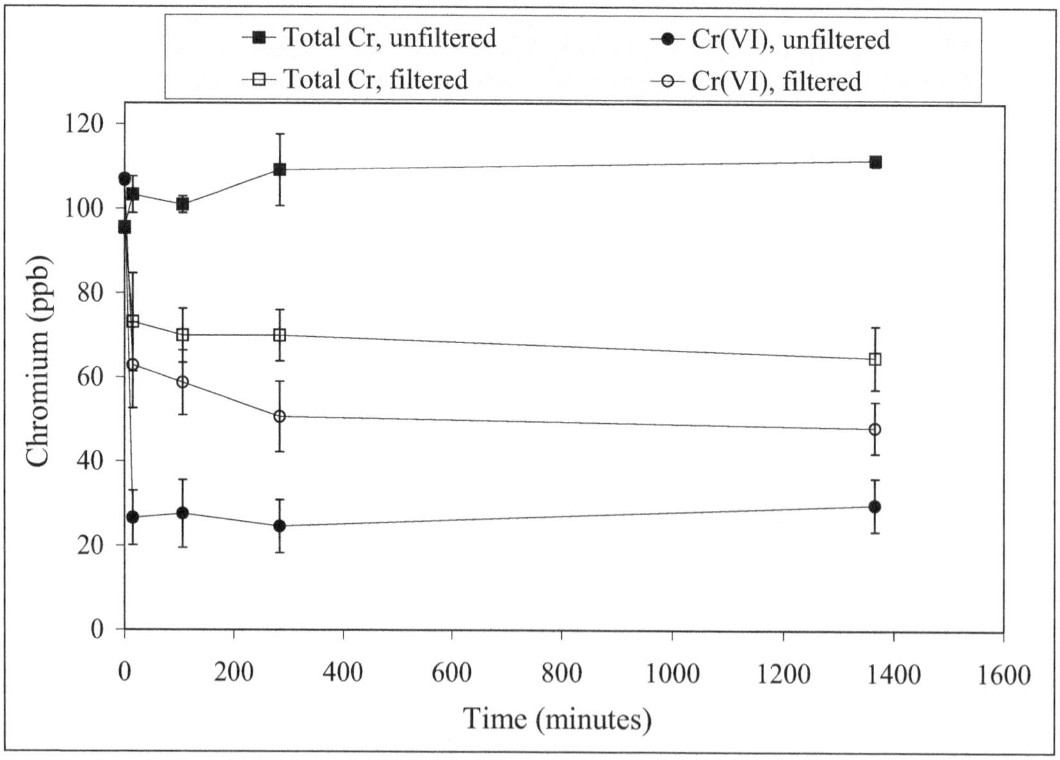

Figure B.15 Reduction of 100 µg/L Cr(VI) by 1.3 mg/L stannous chloride in Glendale water at pH 7

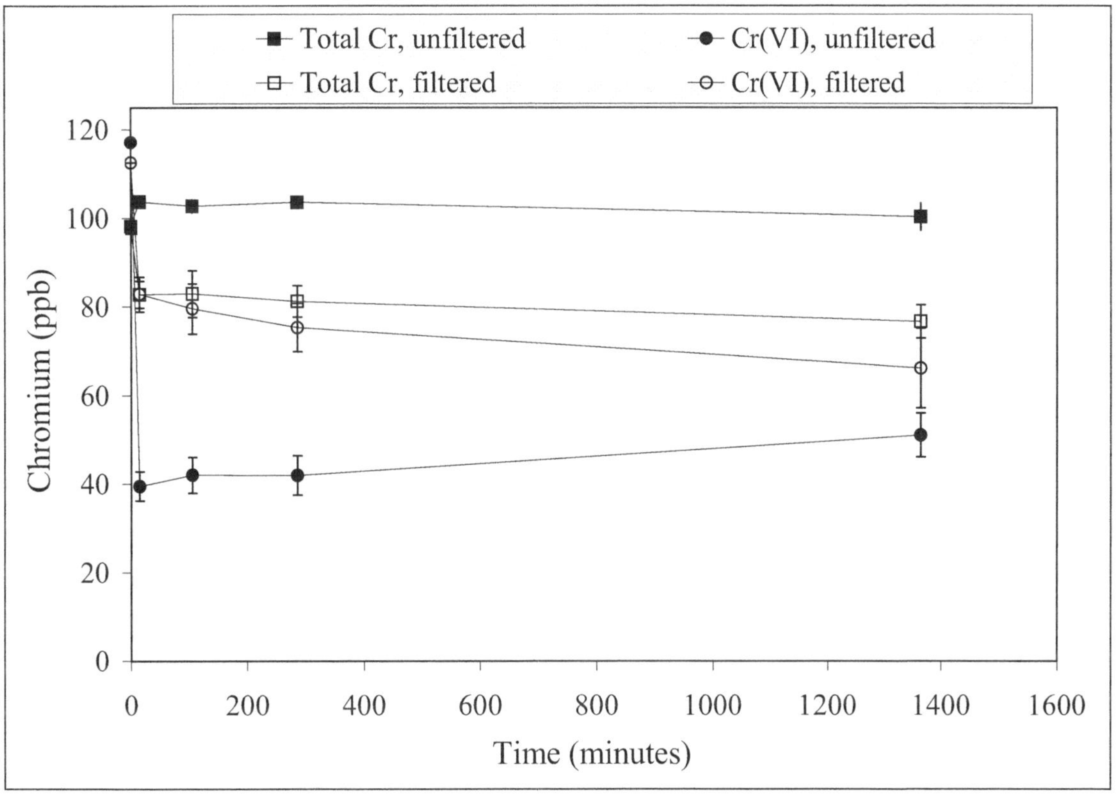

Figure B.16 Reduction of 100 μg/L Cr(VI) by 1.3 mg/L stannous chloride in Glendale water at pH 9

REFERENCES

Agency for Toxic Substances and Disease Registry. 1993. *Toxicological Profile for Chromium.* Washington, D.C.: Department of Health Services, U.S. Public Health Service.

APHA, AWWA and WEF (American Public Health Association, American Water Works Association, and Water Environment Federation). 1998. *Standard Methods for the Examination of Water and Wastewater.* 20th ed. Washington, D.C.: APHA.

Baes, C. F., Jr., and R. E. Mesmer. 1976. The Hydrolysis of Cations. John Wiley and Sons, New York, New York.

Bahowick, S., D. Dobie, and G. Kumamoto. 1996. Ion-Exchange Resin for Removing Hexavalent Chromium From Groundwater at Treatment Facilities. Lawrence, Calif.: Lawrence Livermore National Laboratory, UCRL-JC-122285 Abs.

Baudry, M.R., A. Bouguen, D. Lucas, and B. Chaufer. 1998. Physico-Chemical Characterization of Proteins by Capillary Electrophoresis. *Jour. Chromatography B.*, 706:23–32.

Baudry, M.R., B. Chaufer, P. Aimar, B. Bariou, and D. Lucas. 2000. Application of a Convection–Diffusion–Electrophoretic Migration Model to Ultrafiltration of Lysozyme at Different pH Values and Ionic Strengths. *Jour. Membrane Science*, 179 (1-2):163–174.

Bellamy, L.J. 1975. *The Infra-Red Spectra of Complex Molecules.* London: Chapman and Hall.

Besselievre, E.B. 1969. *The Treatment of Industrial Wastes.* New York: McGraw-Hill.

Beverly, S., S. Seal, and S. Hong. 2000. Identification of Surface Chemical Functional Groups in Reverse Osmosis Membranes: An X-ray Photoelectron Spectroscopy Study. *Jour. Vacuum Science and Technology*, 18(4):1107–1113.

Buerge, I.J., and S.J. Hug. 1997. Kinetics and pH Dependence of Chromium(VI) Reduction by Iron(II). *Environmental Science and Technology.*, 31(5):1426–1432.

Butler, L.N. 1967. *Ionic Equilibrium: A Mathematical Approach.* New York: Addison-Wesley.

Calder, L.M. 1988. Chromium Contamination of Groundwater. In *Chromium in the Natural and Human Environments*. Edited by J.O. Nriagu and E. Nieboer. New York: John Wiley and Sons.

Childress, A., and M. Elimelech. 1996. Effect of Solution Chemistry on the Surface Charge of Polymeric Reverse Osmosis and Nanofiltration Membranes. *Jour. Membrane Science*, 119(2):253–268.

Clifford, D. 1990. Ion Exchange and Inorganic Adsorption. In *Water Quality and Treatment: A Handbook of Community Water Supplies.* Edited by F. Pontius. New York: American Water Works Association, McGraw-Hill, Inc.

Clifford, D., and J. Man Chau. 1987. The Fate of Chromium (III) in Chlorinated Water. EPA/600/2-87/100. Washington, D.C.: U.S. Environmental Protection Agency.

Cox, A.G., and C.W. McLeod. 1992. Field Sampling Technique for Speciation of Inorganic Chromium in River Water. Mikrochimica Acta, 109:161–164.

Eary, L.E., and D. Rai. 1988. Chromate Removal From Aqueous Wastes by Reduction With Ferrous Iron. *Environmental Science and Technology,* 22(8):972–977.

Elimelech, M., W.H. Chen, and J.J. Waypa. 1994. Measuring the Zeta (Electrokinetic) Potential of Reverse Osmosis Membranes by a Streaming Potential Analyzer. *Desalination,* 95(3):269–286.

Elimelech, M., and C.R. O'Melia, 1990. Effect of Electrolyte Type on the Electrophoretic Mobility of Polystyrene Latex Colloids. *Colloids and Surfaces*, 44(3):165–177.

Flegal, R. et al. 2001. Scientific Review of Toxicological and Human Health Issues Related to the Development of a Public Health Goal for Chromium.

Glater, J., J.W. McCutchan, S.B. McCray, and M.R. Zachariah. 1981. The Effect of Halogens on the Performance and Durability of Reverse Osmosis Membranes. American Chemical Society Symposium.

Han, I., M.A. Schlautman, and B. Batchelor. 2000. Removal of Hexavalent Chromium From Groundwater by Granular Activated Carbon. *Water Environment Research*, 72(1):29–39.

Her, N. 2002. Identification and Characterization of Foulants and Scalants on NF Membranes. Ph.D. diss., University of Colorado, Boulder, Colo.

Her, N., G. Amy, and C. Jarusutthirak. 2000. Seasonal Variations of Nanofiltration (NF) Foulants: Identification and Control. *Desalination*, 132(12):143–160.

Hindin, E., G.H. Dunston, and R.J. Bennett. 1968. Water Reclamation by Reverse Osmosis. Bulletin 310. Pullman, Wash.: Washington State University Technical Extension Service.

Hodgman, C.D., R.C. Weast, and S.M. Selby, eds. 1961. *Handbook of Chemistry and Physics*, 43rd ed. Cleveland, Ohio: Chemical Rubber Publishing Company.

Hung, C.P., and M.H. Wu. 1977. Removal of Chromium (VI) From Dilute Aqueous Solution by Activated Carbon, *Water Research*, 11(8):673.

Jakobsen, K., and R. Laska. 1977. Advanced Treatment Methods for Electroplating Wasters. *Pollution Engineering*, 8:42–46

Kaczynski, S.E., and R.J. Kieber. 1993. Aqueous Trivalent Chromium Photoproduction in Natural Water. *Environmental Science and Technology*, 27(8):1572–1576.

Kedem, O., and A. Katchalsky. 1958. Thermodynamic Analysis of the Permeability of Biological Membranes to Non-Elelctrolytes. Biophysica Acta, 27:229.

Kimbrough, D.E. 2002. A Review of the Carcinogenicity of Chromate in Drinking Water. In *Proc. of the AWWA Annual Conference*. Denver, Colo.: AWWA.

Linstedt, K.D., C.P. Huack, and J.T. O'Connor. 1971. Trace Element Removals in Advanced Wastewater Treatment Processes. *Jour. WPCF*, 43(7):1507–1513.

Machle, W., and F. Gregorious. 1942. Cancer of the Respiratory System in the United States Chromate-Producing Industry. *Public Health Report*, 63:114–1127.

Mackay, D.M., and J.A. Cherry. 1989. Groundwater Contamination: Pump-and-Treat Remediation. *Environmental Science and Technology*, 23(6):630–636.

McGuire Environmental Consultants. 2001. Regulatory Assessment and Control Alternatives for Hexavalent Chromium. Report prepared for the City of Glendale, Calif. Santa Monica, Calif.: McGuire Environmental Consultants.

McGuire Environmental Consultants. 2002. Interim Treatment of Cr6 in Glendale's Water Treatment Plant Supply. Report prepared for the City of Glendale, Calif. Santa Monica, Calif.: McGuire Environmental Consultants.

Mixon, F.O. 1973. Removal of Toxic Metals From Water by Reverse Osmosis. R&D Progress Report 889, Saline Water. Washington, D.C.: U.S. Department of the Interior.

Mulder, M. 1991. *Basic Principles of Membrane Technology*. Dordrecht, The Netherlands: Kluwer Academic Publishers.

Nriagu, J. 1988. Production and Uses of Chromium. In *Chromium in the Natural and Human Environments*. Edited by J.O. Nriagu and E. Nieboer. New York: John Wiley and Sons.

Nriagu, J.O., and Nieboer, E. (Eds.) 1988. *Chromium in the Natural and Human Environments*. New York: John Wiley and Sons.

OEHHA 1999. Public Health Goal for Chromium in Drinking Water. California Environmental Protection Agency.

Palmer, C.D., and P.R. Wittbrodt. 1991. Processes Affecting the Remediation of Chromium-Contaminated Sites. *Environmental Health Perspectives*, 92:25–40

Patterson, J. 1985. *Industrial Wastewater Treatment Technology*, 2nd ed. Boston, Mass.: Butterworth.

Philipot, J.M., F. Chaffange, and J. Sibony. 1984. Hexavalent Chromium Removal From Drinking Water. *Water Science and Technology*, 185(17):1121–1132.

Porter, M.C. 1972. Concentration Polarization with Membrane Ultrafiltration. *Industrial and Engineering Chemistry Product Research and Development*, 11(3):234.

Robertson, F.N. 1975. Hexavalent Chromium in the Groundwater in Paradise Valley, Arizona. *Ground Water*, 13:516-527.

Sato, Y., M. Kang, T. Kamei, and Y. Magara. 2002. Performance of Nanofiltration for Arsenic Removal. *Water Research*, 36(13):3371–3377.

See, D.M., and R.E. White. 1999. A Simple Method for Determining Differential Diffusion Coefficients From Aqueous Diaphragm Cell Data at Temperatures Below 0°C. *Jour. Electrochemical Society*, 146(2):677.

Sengupta, A., D. Clifford, and S. Subramonian. 1986. Chromate Ion-Exchange Process at Alkaline pH. *Water Research*, 20(9):77–84.

Sorg, T.J. 1979. Treatment Technology to Meet the Interim Primary Drinking Water Regulation for Organics: Part 4. *Jour. AWWA*, 71(8):454–460.

Tong, J.Y., and E.L. King. 1953. A Spectrophotometric Investigation of the Equilibria Existing in Acidic Solutions of Chromium(VI). *Jour. American Chemical Society*, 75:1902–1906.

Toyama, K., K. Osuga, and S. Maruyama. 1978. On the Formation of Hexavalent Chromium From Chlorine Residue. *Niigata Rikagaku*, 4:38–39.

Tuwiner, S.B. 1962. *Diffusion and Membrane Technology*. New York: Reinhold.

Ulmer, N.S. 1986. Effect of Chlorination on Chromium Speciation in Tap Water, USEPA, EPA/600/M-86/015.

U.S. Environmental Protection Agency. 1996a. Recent Developments for In Situ Treatment of Metals-Contaminated Soil. Draft report. Washington, D.C.: USEPA Office of Solid Waste and Emergency Responses.

U.S. Environmental Protection Agency. 1996b. Method 1636: Determination of Hexavalent Chromium by Ion Chromatography. EPA-821/R-96/003. Washington, D.C.: USEPA Office of Water.

U.S. Environmental Protection Agency. 1999. Method 200.8, Revision 5.5: Determination of Trace Elements in Waters and Wastes by Inductively Coupled Plasma–Mass Spectrometry. EPA-821/R-99/017. Washington D.C.: USEPA Office of Water.

U.S. Environmental Protection Agency. 2000. Definition and Procedure for the Determination of the Method Detection Limit—Revision 1.11 (7-1-00 ed.). 40 CFR 1(136 Part B):310–313.

U.S. Public Health Service. 1953. Health of Workers in the Chromate Industry. USPHS Publication 192, Washington, D.C.: U.S. Government Printing Office.

Vincent, J.B. 2000. The Biochemistry of Chromium. *Jour. of Nutrition*, 130:715–718.

Weast, R.C., M.J. Astle, and W.H. Beyer, eds. 1988. CRC Handbook of Chemistry and Physics, 69th ed. Boca Raton, Fla.: Chemical Rubber Company.

Yoon, Y. 2001. Rejection of Perchlorate by Reverse Osmosis (RO), Nanofiltration (NF), and Ultrafiltration (UF) Membranes: Mechanisms and Modeling. Ph.D. diss., Boulder, Colo.: University of Colorado at Boulder.

Yoshida, H., K. Kamegawa, and S. Arita. 1977. Adsorption of Heavy Metal Ions on Activated Carbon: Adsorption and Reduction of Chromium (VI) on Activated Carbon. *Nippon Kagaku Kaishi*, 3:387.

ABBREVIATIONS

Ar	argon
As(III)	arsenic (III), arsenite
As(V)	arsenic (V), arsenate
ATR–FTIR	attenuated total reflection–Fourier transform infrared
°C	degrees Celsius
C	carbon
$CaCl_2$	calcium chloride
$CaCO_3$	calcium carbonate
Cl	chloride
Cl_2	chlorine
ClO_4^-	perchlorate
CO_3	carbonate
Cr	chromium
Cr(III)	trivalent chromium
Cr(VI)	hexavalent chromium
CrO_4^{2-}	chromate
$Cr_2O_7^{2-}$	dichromate
$Cr(OH)_3$	chromium (III), hydroxide
CRW	Colorado River water
DI	deionized
DHS	(California) Department of Health Services
DOC	dissolved organic carbon
EBCT	empty bed contact time
ED–EDR	electrodialysis–electrodialysis reversal
Fe	iron
Fe(0)	zerovalent iron
Fe(II)	ferrous iron, iron (II)

Fe(III)	ferric iron, iron (III)
$FeCr_2O_4$	chromite (p. 4)
FTIR	Fourier transform infrared
GAC	granular activated carbon
Ge	germanium
GFH	granular ferric hydroxide
H_2O_2	hydrogen peroxide
HCl	hydrochloric acid
HCO_3	bicarbonate
$HCrO_4^-/CrO_4^{2-}$	chromate
HDPE	high-density polyethylene
HNO_3	nitric acid
hr	hour, hours
IC	ion chromatography
ICP–MS	inductively coupled plasma–mass spectrometer
KCl	potassium chloride
$KMnO_4$	potassium permanganate
KOH	potassium hydroxide
kPa	kilopascal
K_2SO_4	potassium sulfate
L	liter
M	molar
MCL	maximum contaminant level
MCLG	maximum contaminant level goal
MDL	method detection limit
meq	milliequivalent
Mg	magnesium

$MgSO_4$	magnesium sulfate
mg/L	milligram per liter
min	minute
mL	milliliter
mM	millimolar
MMCO	molecular mass cutoff
Mn	manganese
MnO_4	permanganate
MΩ	megaohm
mS/m	millisiemens per meter
µg/L	microgram per liter
µL	microliter
MW	molecular weight
MWCO	molecular weight cutoff
N	Newton
Na	sodium
NaCl	sodium chloride
Na_2CO_3	sodium carbonate
$NaNO_3$	sodium nitrate
NaOCl	sodium hypochlorite
NaOH	sodium hydroxide
$Na_2S.9H_2O$	sodium sulfide nonahydrate
$Na_2SiO_3.9H_2O$	sodium silicate nonahydrate
Na_2SO_3	sodium sufite
NaOH	sodium hydroxide
$(NH_4)_2SO_4$	ammonium sulfate
NOM	natural organic matter
NWIS	(USGS) National Water Information System
OEHHA	Office of Environmental Health Hazards Assessment

PAC	powdered activated carbon
pE	negative logarithm of apparent electron activity; indicator of oxidation–reduction (redox) potential
PHG	public health goal
redox	reduction–oxidation
rpm	revolutions per minute
S	sulfur
S^{-2}	sulfide
SBA	strong base anion
SiO_2	silica
SO_3	sulfite
SO_4	sulfate
$SnCl_2$	stannous chloride
$SnCl_2.2H_2O$	stannous chloride dihydrate
TCLP	Toxic Characteristic Leaching Protocol
TDS	total dissolved solids
TOC	total organic carbon
USEPA	U.S. Environmental Protection Agency
USPHS	U.S. Public Health Service
UV_{254}	ultraviolet absorbance at 254 nanometers
WET	Waste Extraction Test

IWA Publishing's authorised EU representative for General Product Safety Regulations is Diane D'Arras, 15 rue Duret, 75116 Paris, France, e-mail: safety@iwap.co.uk.

Printed and bound by CPI Group (UK) Ltd, Croydon, CR0 4YY
17/09/2026
02218793-0004